Power Plant Stability, Capacitors, and Grounding

Power Plant Stability, Capacitors, and Grounding
Numerical Solutions

Orlando N. Acosta, MSEE

New York   Chicago   San Francisco
Lisbon   London   Madrid   Mexico City
Milan   New Delhi   San Juan
Seoul   Singapore   Sydney   Toronto

The **McGraw·Hill** Companies

Copyright © 2012 by The McGraw-Hill Companies, Inc. All rights reserved. Printed in the United States of America. Except as permitted under the United States Copyright Act of 1976, no part of this publication may be reproduced or distributed in any form or by any means, or stored in a data base or retrieval system, without the prior written permission of the publisher.

1 2 3 4 5 6 7 8 9 0   DOC/DOC   1 8 7 6 5 4 3 2

ISBN 978-0-07-180008-2
MHID 0-07-180008-5

Sponsoring Editor
Michael Penn

Editing Supervisor
Stephen M. Smith

Production Supervisor
Richard C. Ruzycka

Acquisitions Coordinator
Bridget L. Thoreson

Project Manager
Rohini Deb,
Cenveo Publisher Services

Copy Editor
Patti Scott

Proofreader
Carol Shields

Indexer
Robert Swanson

Art Director, Cover
Jeff Weeks

Composition
Cenveo Publisher Services

Printed and bound by RR Donnelley.

McGraw-Hill books are available at special quantity discounts to use as premiums and sales promotions, or for use in corporate training programs. To contact a representative, please e-mail us at bulksales@mcgraw-hill.com.

This book is printed on acid-free paper.

Information contained in this work has been obtained by The McGraw-Hill Companies, Inc. ("McGraw-Hill") from sources believed to be reliable. However, neither McGraw-Hill nor its authors guarantee the accuracy or completeness of any information published herein, and neither McGraw-Hill nor its authors shall be responsible for any errors, omissions, or damages arising out of use of this information. This work is published with the understanding that McGraw-Hill and its authors are supplying information but are not attempting to render engineering or other professional services. If such services are required, the assistance of an appropriate professional should be sought.

About the Author
Orlando N. Acosta, MSEE, has worked as an electrical engineer and consultant for Sunbelt Energy Systems, Naval Sea Systems Command, Parsons & Whittemore, Catalytic Inc., Chrysler Corporation Space Division, Sperry Rand Corporation, ITE Circuit Breaker, and CENO Utility.

Contents

Preface xi

1 **Power System Basic Knowledge** 1
   1.1 Three-Phase Balanced Circuits 1
   1.2 Reduction of Electrical Networks 2
   1.3 Per-Unit Quantities 4
   1.4 MVA Method of Short Circuit
        Calculation 6
   1.5 Short Circuit MVA Combination Rules 7
   1.6 Iron Core Saturation 12

2 **Power Systems Stability** 17
   2.1 Introduction 17
   2.2 Classical Model 18
   2.3 Power Flow from Generator to Motor 20
   2.4 Steady-State Stability 25
   2.5 Brief Summary of Rotational Dynamics 26
   2.6 The Swing Equation 30
   2.7 Synchronizing Power Coefficient 32
   2.8 Natural Frequency of Oscillation 35
   2.9 Equal-Area Criterion of Stability 36
   2.10 Generator-Infinity Bus Network 39
   2.11 Introduction to Stability of Multimachine
         Power Systems 40
   2.12 Coherent Machines 41
   2.13 Modeling of Multimachine Power
         Systems 42
   2.14 Power Flow in a Multimachine Network ... 43
   2.15 Network Reduction Techniques 44

3 **Transient Stability Problem in a Simple Electrical
Network** 49
   3.1 Stability Problem 49
   3.2 Network Reduction 50
   3.3 Electric Power Transmitted 53
   3.4 Power Transmitted Before, During, and After
        Fault Conditions 55
   3.5 Swing Equation 56
   3.6 Numerical Solver 58

Contents

4 Transient Stability Problem in a Multimachine Network **65**
   4.1 Minimum Data Necessary to Do a Transient Stability Study 68
   4.2 Converting Electrical Loads to Equivalent Admittances 71
   4.3 Load Flow during Normal Operation 73
   4.4 Initial Power Angle Computation 87
   4.5 Network Configuration during the Fault at F1 89
   4.6 Numerical Solution of the Swing Equation 96

5 High-Voltage AC Capacitors **103**
   5.1 Introduction 103
   5.2 Capacitor Steady-State Equations 105
   5.3 Basic Capacitor Connections 105
   5.4 Reactive Power Compensation 107
   5.5 Series-Connected Capacitor Banks 108
   5.6 Shunt-Connected Capacitor Banks 110
   5.7 AC Voltage Suddenly Applied To or Removed From an RLC Series Circuit 112

6 Substation Grounding **127**
   6.1 Background 127
   6.2 Approaches to Grid Design 127
   6.3 Generally Accepted Assumptions 128
   6.4 Separated Ground Rods 129
   6.5 Substation Fences 129

7 Dangerous Electric Currents **131**
   7.1 Background 131
   7.2 Magnitude and Frequency 132
   7.3 Duration and Current Path 133
   7.4 Electrical Substation Grounding 137
   7.5 Important Voltage Gradient Definitions 138

8 Ground Grid Preliminary Design **139**
   8.1 Background 139
   8.2 Single-Rod Electrodes 140
   8.3 Ground Mat Resistance to Earth, Approximated Formulas 142
   8.4 Ground Mat Conductor Corrosion 143
   8.5 Grid Conductor Size 145
   8.6 Gradient Control 148
   8.7 Example of Preliminary Grid Design 152

Contents

9 **Principles of Ground Mat Design** **159**
   9.1 Introduction 159
   9.2 Potential Created by a Point Current Source 161
   9.3 Potential at a Point inside Earth Created by Current Leaking to Earth from a Segment of a Grid Conductor 163
   9.4 Mutual Resistance between Two Conductor Segments 167
   9.5 Self-Resistance 174

10 **Ground Mat Design with Nonuniform Current Distribution** **177**
   10.1 Introduction 177
   10.2 Grid Current Distribution during a Fault to Ground 177
   10.3 Computations with Nonuniform Current Distribution in Small Square Grid 180
   10.4 Ground Grid Buried in Top Layer of Two-Layer Earth Model 203
   10.5 Ground Grid Buried in Bottom Layer of Two-Layer Earth Model 206

Bibliography 209

Index 211

Preface

The mysterious blackouts that sometimes occur in large parts of the North American continent present a real conundrum to utility companies, electrical equipment manufacturers, and government regulatory agencies. The electrical engineering expertise required to design, install, maintain, and operate the vast network of interconnected electrical power systems existing in the United States and Canada is always increasing, and the tempo has consistently accelerated since the 1960s. Fortunately, new powerful tools, like digital computers and the Internet, facilitate the study and solution of engineering problems. Although this book was written using MathCad 14 software, the reader does not need to have the program. The numerical solution of differential equations is used throughout the book because it is an important computational tool for solving complex problems like stability of generating power systems.

Power electrical engineers involved in the design, installation, operation, or maintenance of electrical equipment will welcome the detailed explanations and the emphasis given to the solution of complicated electrical problems. In all the examples solved, the intent is to explain the method of solution rather than provide a quick answer.

This book has been written fundamentally to help practicing engineers solve intricate problems without having to relearn their university professors' mathematical jargon that they may have forgotten a long time ago. To that effect, the material is presented, as much as possible, using elementary mathematical tools. Furthermore, the main goal of this book is to be easily understood and with that purpose the material can at times be repetitive. However, the solution methods described and implemented are valid, precise, and up-to-date.

Practicing engineers will quickly find out that studies based on the methods presented in this book will be welcome and easily understood by people attending engineering meetings, and that they will not hesitate in providing suggestions based on commonsense engineering practice.

Chapter 1 is a brief review of fundamental electrical engineering concepts. Chapters 2, 3, and 4 present and discuss the stability of power systems. In them, problems and examples are worked out to the last detail and discussed. The solution conclusions are explained with the help of graphics.

Chapter 5 treats high-voltage capacitors, with special emphasis on the current and voltage oscillations that they can introduce in power systems when charging or discharging.

Chapters 6 to 10 cover, in great detail, electrical substation grounding systems, including ground grid design with nonuniform current distribution.

Although I have not attempted to give specific credit at each point, I have used the books listed in the bibliography as sources of data and information, and my style of writing has been influenced by them.

Orlando N. Acosta, MSEE

Power Plant Stability, Capacitors, and Grounding

CHAPTER 1
Power System Basic Knowledge

The material contained in this chapter is only a small fraction of the basic knowledge that every electrical engineer should have. Any power system electrical engineer should know the methods to design a capacitor bank, to make a load study, or to predict the response of a given power system to transient disturbances. Fortunately, engineers now have very powerful analytical tools downloaded into their computers that allow them to conduct studies with easy and excellent results and to display the results in two or three dimensions and wonderful colors.

1.1 Three-Phase Balanced Circuits

Polyphase alternating-current (AC) systems for power generation, transmission, and distribution are almost exclusively three-phase systems. A balanced three-phase circuit has identical line and load impedances in all phases and the voltages impressed on the impedances are equal in magnitude and 120° out of phase with one another. In this kind of circuit the calculations can be made on a per-phase basis and using single-phase circuit analysis concepts.

A power system can fail in the short circuit mode or in the open-phase mode; the open-phase type of failure occurs very seldom and is not treated in this book. For three-phase systems the possible types of short circuit failures are three-phase, phase to phase, phase to ground, and two-phase to ground. In utility and industrial power systems design, the three-phase short circuit is by far the most important type, not because it occurs more often, but because it can be compared with the short circuit rating of electrical devices, and because it usually produces the largest short circuit current. Only the phase-to-ground short circuits can produce, in rare cases, a larger current than the three-phase short circuit, up to 125 percent larger. In the event of three-phase short circuits, the three-phase balanced power system remains balanced, and therefore the short circuit computations can be

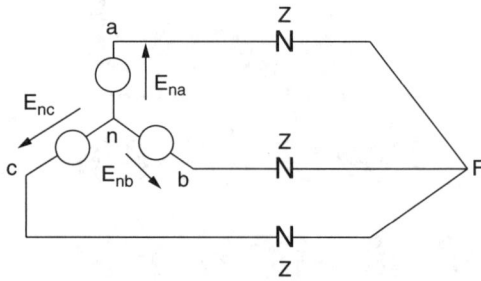
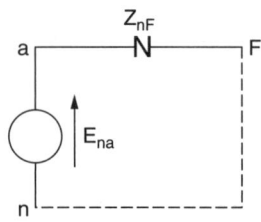

FIGURE 1-1 Diagram of a three-phase short circuit in a balanced three-phase circuit with one voltage source.

FIGURE 1-2 Equivalent single-line diagram.

made on a per-phase basis. Furthermore, the potential at the fault location is equal to the potential of the system neutral. This is true for all balanced three-phase power systems, with grounded or ungrounded neutral, regardless of the number of wires; i.e., whether the system has three or four wires makes no difference. In a balanced three-phase, four-wire circuit, the current in the neutral is always zero, even during a three-phase short circuit event. This type of short circuit does not produce voltage distortion or current unbalance. Figure 1-1 illustrates a three-phase short circuit in a balanced three-phase system with one voltage source, and Fig. 1-2 shows the equivalent single-line diagram. Real circuits are usually not balanced, because any asymmetry in the physical location of the conductors or unbalanced loading will make the system unbalanced. However, it is common practice in analytical studies to consider the system symmetrical and balanced.

1.2 Reduction of Electrical Networks

There are several theorems that are essential to the reduction techniques of electrical networks: Thévenin's theorem, Norton's theorem, and the superposition theorem. Thévenin's theorem is applicable to linear, two-terminal circuits and can be enunciated as follows: If a load with impedance Z_L is connected between any two points of an energized circuit, then the resulting current I_L through this impedance is the potential difference E_p between these points prior to the connection, divided by the sum of the connected impedance Z_L and the impedance Z_{in}, which is the impedance of the rest of the circuit looking back into the circuit from the points across which the impedance Z_L is connected. To evaluate Z_{in}, all sources of emf must be assumed to be zero and replaced by their internal impedances.

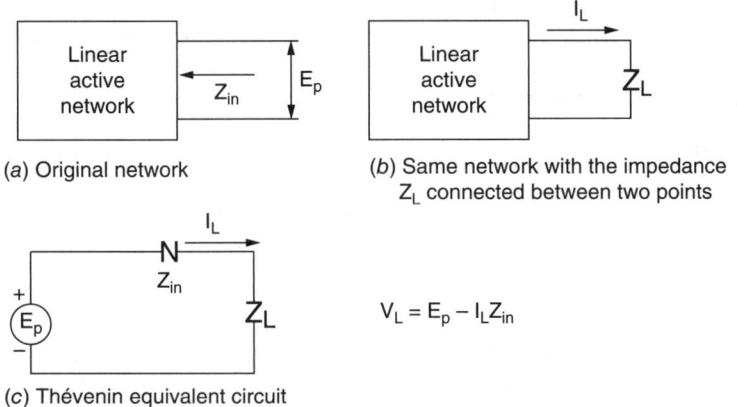

FIGURE 1-3 Diagrams illustrating Thévenin's theorem.

Figure 1-3 illustrates the application of Thévenin's theorem to an active two-terminal network.

Norton's theorem is very similar to Thévenin's theorem. It states that with respect to any pair of terminals of a linear and active network, the network can be substituted by an equivalent circuit that has a single current source. This source is in parallel with an impedance, equal to the impedance of the network looking back into the network from the specified pair of terminals, provided that any emf existing in the network is short-circuited and replaced by its internal impedance and that any current source is opened. This parallel impedance is equal to the series impedance of the Thévenin equivalent circuit. The constant current, delivered by the constant current source, is equal to the open-circuit voltage across the specified pair of terminals divided by the parallel impedance. The voltage of the current source is equal to the prior connection voltage of the Thévenin equivalent circuit. Figure 1-4 illustrates the application of Norton's theorem to an active two-terminal network. Assume that a passive load is connected across terminals a and b of the network depicted in Fig. 1-4a. This implies that any emf existing in the

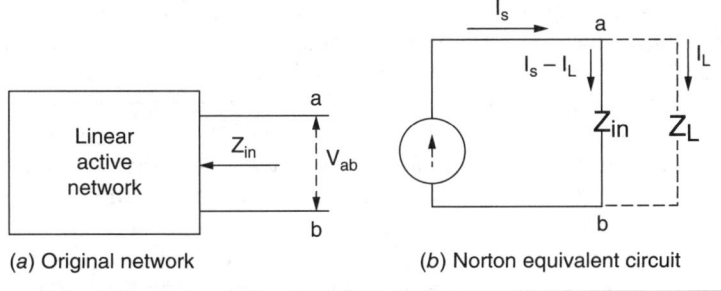

FIGURE 1-4 Diagrams illustrating Norton's theorem.

4 Chapter One

load is short-circuited and any current source opened. Let us compare the Thévenin equivalent circuit with the Norton equivalent circuit.

Thévenin's equation: $V_L = E_p - I_L \times Z_{in}$ See Fig. 1-3.

Norton's equations:

$$V_L = (I_S - I_L) \times Z_{in} = I_S \times Z_{in} - I_L \times Z_{in} \qquad I_s = \frac{V_{ab}}{Z_{in}} \qquad I_s \times Z_{in} = V_{ab}$$

$$V_L = V_{ab} - I_L \times Z_{in}$$

When using Thévenin's theorem in short circuit computations, it is a common practice to consider the impedance of the fault to be zero. The system nominal line to neutral voltage is used instead of the pre-fault voltage to neutral at fault location; besides the resistive portion of looking in impedance is usually neglected.

In both the Thévenin and Norton equations, V_L is the voltage across the connected load and I_L is the load current. Thévenin's equation is equal to Norton's equation because by definition $E_p = V_{ab}$.

Superposition Theorem

The superposition theorem states that in a network containing several emf sources each may be independently considered, and that the total current in any branch of the network produced by all the emf's is the phasor addition of the individual currents produced by each emf acting separately. To compute the current due to any one of the emf's, all the other sources of emf must be assumed to be zero and replaced by their internal impedances. These three theorems, together with the fact that three-phase balanced circuits can be treated as single-phase or two-terminal networks, make the analysis and computations much easier.

1.3 Per-Unit Quantities

The per-unit method of expressing electrical quantities is based on ohm's law. This law relates three variables with one equation; therefore only two variables can be arbitrarily selected as independent. In the power industry, the most common selection is three-phase megavolt-amperes (MVA) and line-to-line kilovolts as the two independent base quantities, and all other quantities are expressed as functions of them. It is necessary to say that the MVA quantity is a combination of two of the three original quantities related by ohm's law.

The per-unit value of a quantity is the ratio of the quantity to its base. The percentage value is the per-unit value times 100. For balanced three-phase systems, the per-unit value of a line-to-neutral voltage with line-to-neutral base is the same as the per-unit value, at the same location, of the line-to-line voltage with line-to-line base. Likewise, the total three-phase MVA is three times the MVA per phase, because the three-phase MVA base is three times the base MVA per phase. Then the per-unit value of the three-phase MVA with a three-phase MVA base is also the same as the per-unit value of the MVA per phase with an MVA per phase base. Symbolically,

$$\frac{V_{LN}}{V_{LNb}} \quad \text{per-unit value of line-to-neutral (LN) voltage}$$

In a balanced three-phase circuit, the per-unit value of the line-to-line (LL) voltage is equal to the per-unit value of the line-to-neutral voltage. Symbolically,

$$\frac{V_{LL}}{V_{LLb}} = \frac{\sqrt{3} \times V_{LN}}{\sqrt{3} \times V_{LNb}} = \frac{V_{LN}}{V_{LNb}}$$

In the megavolt-amperes case, we have

$$\frac{MVA}{MVA_b} = \frac{3 \times MVA_{ph}}{3 \times MVA_{bph}} = \frac{MVA_{ph}}{MVA_{bph}} \quad \begin{array}{l}\text{per-unit value of three-}\\ \text{phase MVA is same as that}\\ \text{of MVA per phase}\end{array}$$

Usually, computations in balanced three-phase circuits are made on a per-phase basis in the same way as calculations are made for single-phase circuits. However, the three-phase megavolt-amperes of a balanced three-phase system are given in terms of the line-to-line voltage and line current by the same equation, regardless of whether the system is wye- or delta-connected. Symbolically,

$$MVA = \sqrt{3} \times kV_{LL} \times kI_L$$

From here on, unless otherwise specified, the notation for base quantities is

MVA_b = base three-phase megavolt-amperes
kV_b = base line-to-line kilovolts
I_b = base line current, in kA
Z_b = base line-to-neutral impedance in ohms

$$MVA_b = \sqrt{3} \times kV_b \times I_b \quad I_b = \frac{MVA_b}{\sqrt{3} \times kV_b} \quad \text{base kiloamperes} \quad (1.1)$$

$$Z_b = \frac{kV_b}{\sqrt{3} \times I_b} = \frac{kV_b \times \sqrt{3} \times kV_b}{\sqrt{3} \times MVA_b} = \frac{(kV_b)^2}{MVA_b} \quad \begin{array}{l}\text{base line to}\\ \text{neutral ohms}\end{array} \quad (1.2)$$

Conversion of Per-Unit Impedance Values from One Base to Another

Per-unit impedance = actual impedance/base impedance. Symbolically:

$$\frac{\Omega}{\Omega_b} \qquad (1.3)$$

Substituting Eq. (1.2) into Eq. (1.3), we obtain

$$\frac{\Omega \times MVA_b}{(kV_b)^2} \qquad \text{per-unit impedance}$$

If we express the per-unit impedance in two different bases, we obtain

$$\frac{\Omega \times MVA_{b1}}{(kV_{b1})^2} = Z_1 \qquad \frac{\Omega \times MVA_{b2}}{(kV_{b2})^2} = Z_2 \qquad \frac{Z_2}{Z_1} = \frac{\Omega \times MVA_{b2} \times (kV_{b1})^2}{\Omega \times MVA_{b1} \times (kV_{b2})^2}$$

$$Z_2 = Z_1 \times \frac{MVA_{b2}}{MVA_{b1}} \times \left(\frac{kV_{b1}}{kV_{b2}}\right)^2 \qquad \begin{array}{l}\text{impedance in base 2 given} \\ \text{impedance in base 1}\end{array} \qquad (1.4)$$

Given the per-unit impedance in base 1, Eq. (1.4) provides the per-unit impedance in base 2.

1.4 MVA Method of Short Circuit Calculation

The MVA method is a different way of resolving short circuit problems that is easier to remember and handle. Although it can be used to resolve any kind of short circuit, in this book the emphasis is on three-phase short circuit (SC) calculations. These calculations are based on Eqs. (1.5) and (1.6).

$$MVA_{sc} = \frac{kV^2}{Z} \qquad (1.5)$$

$$MVA_{sc} = \frac{MVA_b}{Z_{pu}} \qquad (1.6)$$

where kV = line-to-line voltage, kV
   Z = line-to-neutral impedance in ohms
   Z_{pu} = per-unit line-to-neutral impedance
   MVA_{sc} = short circuit MVA

Equation (1.5) is mainly applied to feeders and busses. Their short circuit MVA is equal to the square of the line-to-line kilovolts divided by the line-to-neutral impedance in ohms. Equation (1.6) is

Power System Basic Knowledge

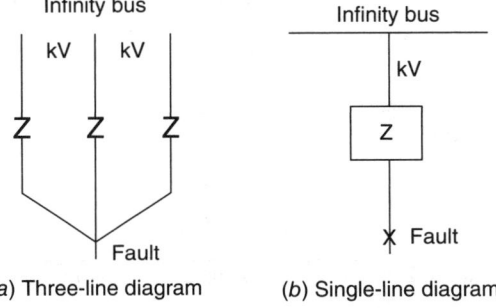

FIGURE 1-5 Illustration to deduce the MVA basic equations.

mainly applied to transformers, motors, and generators; their short circuit MVA is equal to the device MVA base (MVA rating) divided by its own per-unit impedance. Equations (1.5) and (1.6) express the maximum possible flow of short circuit MVA through any given component, which only happens when the component is fed from an infinity bus. These equations are phasor equations. However, they are much easier to apply when only the magnitudes of the phasors are used.

The short circuit in Fig. 1-5 is calculated as follows:

$$I_{sc} = \frac{kV}{\sqrt{3} \times Z} \qquad MVA_{sc} = \sqrt{3} \times kV \times I_{sc} = \sqrt{3} \times kV \times \frac{kV}{\sqrt{3} \times Z} = \frac{kV^2}{Z}$$

same as Eq. (1.5)

Working with Eqs. (1.5) and (1.2) and knowing that by definition $kV = kV_b$, we get

$$MVA_{sc} = \frac{kV^2}{Z_b \times Z_{pu}} = \frac{(kV_b)^2}{Z_b \times Z_{pu}} = \frac{Z_b \times MVA_b}{Z_b \times Z_{pu}} = \frac{MVA_b}{Z_{pu}}$$

same as Eq. (1.6)

where MVA_b and Z_{pu} are part of the device ratings. The maximum possible short circuit MVA through any device is equal to its MVA rating divided by its own per-unit impedance.

1.5 Short Circuit MVA Combination Rules

In the MVA method, the contribution of every device to the total short circuit MVA is computed separately as the maximum possible for that device and then assembled together as for the following rules.

Chapter One

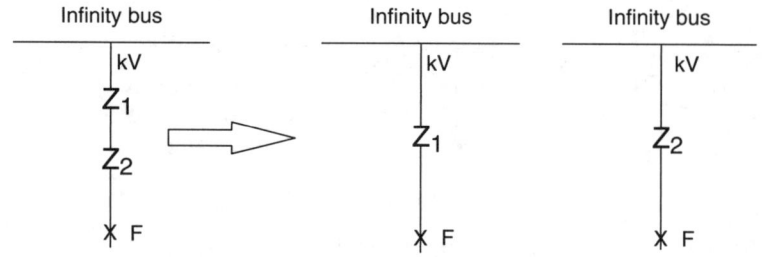

FIGURE 1-6 Diagram illustrating the combination rules for components in series.

Components in Series

The components are separated, and each is considered supplied from an infinity bus and with a fault in its output terminals, as shown in Fig. 1-6. The short circuit MVA flow through each component is calculated using Eq. (1.5). Symbolically,

$$MVA_{sc} = \frac{kV^2}{Z} = kV^2 \times Y \quad MVA_{sc1} = kV^2 \times Y_1 \quad MVA_{sc2} = kV^2 \times Y_2$$

The equivalent circuit of Fig. 1-6 is shown in Fig. 1-7. The equivalent impedance of a circuit with components in series is the sum of the components' impedances. Here is the phasor addition of two components' in series:

$$Z_e = Z_1 + Z_2 \qquad \frac{1}{Y_e} = \frac{1}{Y_1} + \frac{1}{Y_2} = \frac{Y_2 + Y_1}{Y_1 \times Y_2} \qquad Y_e = \frac{Y_1 \times Y_2}{Y_1 + Y_2}$$

where Z_e and Y_e are the impedance and admittance, respectively, of the equivalent circuit. The short circuit MVA of the equivalent circuit

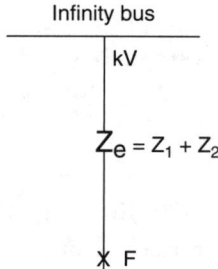

FIGURE 1-7 Equivalent circuit of a circuit with two components in series.

of two components in series, as a function of the short circuit MVA of the individual components, is determined as follows:

$$MVA_{sce} = kV^2 \times Y_e = kV^2 \times \frac{Y_1 \times Y_2}{Y_1 + Y_2} = \frac{(kV^2 \times Y_1)(kV^2 \times Y_2)}{(kV^2 \times Y_1) + (kV^2 \times Y_2)}$$

$$= \frac{MVA_{sc1} \times MVA_{sc2}}{MVA_{sc1} + MVA_{sc2}} \quad (1.7)$$

Equation (1.7) gives the short circuit MVA of two components in series.

Rule

For components connected in series, combine the components' short circuit MVAs using the same procedure as that to combine resistances in parallel. The short circuit MVAs in Eq. (1.7) cannot be summed arithmetically because their angular displacements are different in general, unless the resistive parts of the impedance of the components are neglected. Keep in mind that the impedance of the circuit determines the value of the angle between the voltage and the current. Also, the angle of the MVA is given by $\tan^{-1} = Q/P$ or VAR/watts.

Components in Parallel

The equivalent impedance and equivalent admittance of the equivalent circuit for two components in parallel are

$$Z_e = \frac{Z_1 \times Z_2}{Z_1 + Z_2} \qquad Y_e = Y_1 + Y_2$$

The short circuit MVA of the equivalent circuit of two components in parallel, as a function of the short circuit MVA of the individual components, is determined as follows:

$$MVA_{sce} = kV^2 \times Y_e = kV^2 \times (Y_1 + Y_2) = kV^2 \times Y_1 + kV^2 \times Y_2$$

$$= MVA_{sc1} + MVA_{sc2} \quad (1.8)$$

Equation (1.8) gives the short circuit MVA of two components in parallel.

Rule

For components connected in parallel, combine the components' short circuit MVAs using the same procedure as that to combine resistances in series. Equation (1.8) can be generalized for any number of components in parallel. As stated before, the short circuit MVAs cannot be summed arithmetically unless the resistive parts of the impedance of the components are neglected.

Chapter One

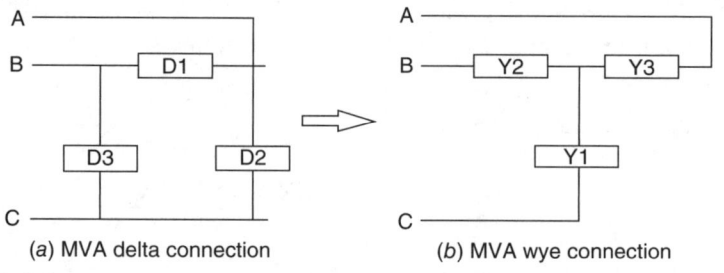

(a) MVA delta connection   (b) MVA wye connection

FIGURE 1-8 Short circuit MVAs delta-to-wye conversion.

Delta-Wye Conversion

A delta configuration of short circuit MVAs can be substituted by an equivalent wye configuration of short circuit MVAs and vice versa. However, the wye-to-delta conversion is not used with the MVA method. The two configurations illustrated in Fig. 1-8 are equivalent if

$$Y_1 = \frac{D1 \times D2 + D2 \times D3 + D3 \times D1}{D1}$$

D's and Y's are short circuit MVAs.

$$Y_2 = \frac{D1 \times D2 + D2 \times D3 + D3 \times D1}{D2}$$

$$Y_3 = \frac{D1 \times D2 + D2 \times D3 + D3 \times D1}{D3} \quad Y = 3 \times D$$

For all equal D components we obtain all equal Y components. Every leg of the equivalent wye is equal to 3D.

The procedure for short circuit current calculation using the MVA method is as follows:

1. Separate the circuit into components.
2. Calculate the maximum MVA flow through each component.
3. Generate the MVA diagram of the circuit, using the values obtained in step 2.
4. Using the combination rules, obtain the total short circuit MVA.
5. Calculate the short circuit rms current.

Power System Basic Knowledge

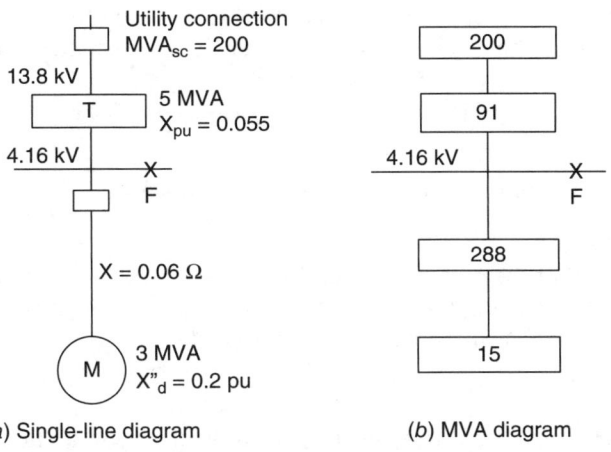

FIGURE 1-9 Three-phase fault solved by the MVA method.

Example 1-1

Find the short circuit current at the marked location of the power system depicted in Fig. 1-9. Applying Eqs. (1.5) and (1.6), we get the MVAs depicted in Fig. 1-9b.

Utility contribution: 200

Transformer contribution: $\dfrac{5}{0.055} = 91$ MVA

Line contribution: $\dfrac{4.16^2}{0.06} = 288$

Motor contribution: $\dfrac{3}{0.2} = 15$

Combining the components' short circuit MVAs, we get the MVAs depicted in Fig. 1-10.

$$\dfrac{200 \times 91}{200 + 91} = 63$$

$$\dfrac{15 \times 288}{15 + 288} = 14$$

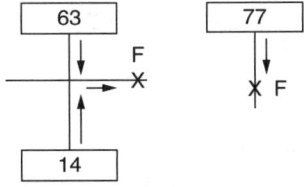

FIGURE 1-10 Short-circuit MVAs combination.

The line reduces the motor contribution. When two components are in series, the result is always smaller than the smaller of the two.

$$\sqrt{3} \times 4.16 \times I_{sc} = 77 \qquad I_{sc} = \frac{77}{\sqrt{3} \times 4.16} = 10.7 \text{ kA}$$

The three-phase short circuit is 10,700 rms A.

Example 1-2

Assume that the motor depicted in Fig. 1-9 drives an air compressor in a manufacturing plant. Find the instantaneous voltage drop at the motor terminals any time the compressor starts.

Motor data:

3000-HP, three-phase, 4000-volts, squirrel cage induction motor
Full load current: 377 A
Locked rotor current: 2070 A

The instantaneous voltage drop in per unit of the initial voltage at the motor terminals could be calculated using the following formula:

$$\Delta V_{pu} = \frac{MVA_{st}}{MVA_{st} + MVA_{sc}}$$

where MVA_{st} = motor starting MVA, assuming initial system voltage remains constant, and MVA_{sc} = available short circuit MVA at the motor terminals without motor contribution.

$$MVA_{st} = \sqrt{3} \times 4.16 \times 2.07 = 15 \qquad MVA_{st} := 15$$

From Fig. 1-9 we calculate the short circuit MVA:

$$\frac{200 \times 91}{291} = 63 \qquad \frac{63 \times 288}{63 + 288} = 52 \qquad MVA_{sc} := 52$$

$$\frac{15}{15+52} = 0.22 \quad \Delta V_{pu} = 0.22 \quad 0.22 \times 4160 = 915.2 \quad \Delta V_M = 915 \text{ volts}$$

$$4160 - 915 = 3245 \qquad V_M = 3245 \text{ volts} \qquad \frac{3245}{4000} = 0.81$$

Voltage will dip to 81 percent of rated voltage. This is no good. Lamps will flicker in the plant any time the compressor starts.

1.6 Iron Core Saturation

The amount of volt-seconds that any device with an iron core can absorb depends on the iron core cross section. A perfect sinusoidal voltage applied to the device winding will produce a perfect hysteresis loop of the magnetic flux density. Any distortion of the half-wave symmetry of the voltage applied will result in a wave shape containing even harmonics or a DC component or both. These harmonics will shift the hysteresis loop in such a way that could saturate (the flux density could not increase more) the iron in one direction or the other. From the saturation point forward, the iron would behave as air. Even harmonics, specially the second, destroy the half-wave symmetry of the incoming voltage, and the effect on the magnetic core is similar to applying a DC bias to the winding. When a device core saturates, it takes very large input current, due to the lack of voltage drop produced by the lack of change in the magnetic flux density, and the device, could burn if the protecting breaker does not open quickly enough.

Let us consider the sinusoidal wave defined by Eq. (1.9) and shown in Fig. 1-11. This is applied to a winding wound on the iron core illustrated in Fig. 1-12.

$$V = V_m \times \sin(\omega t) \tag{1.9}$$

$f := 60 \; \dfrac{\text{cycle}}{\text{sec}} \qquad \omega := 2\pi f = 377 \; \dfrac{\text{rad}}{\text{sec}} \qquad T := \dfrac{1}{60} = 0.017 \; \dfrac{\text{sec}}{\text{cycle}}$

$f := 50 \qquad \omega_1 := 2\pi f_1 = 314 \qquad T_1 := \dfrac{1}{50} = 0.02$

$f_2 := 400 \qquad \omega_2 := 2\pi f_2 = 2513 \qquad T_2 := \dfrac{1}{400} = 0.0025$

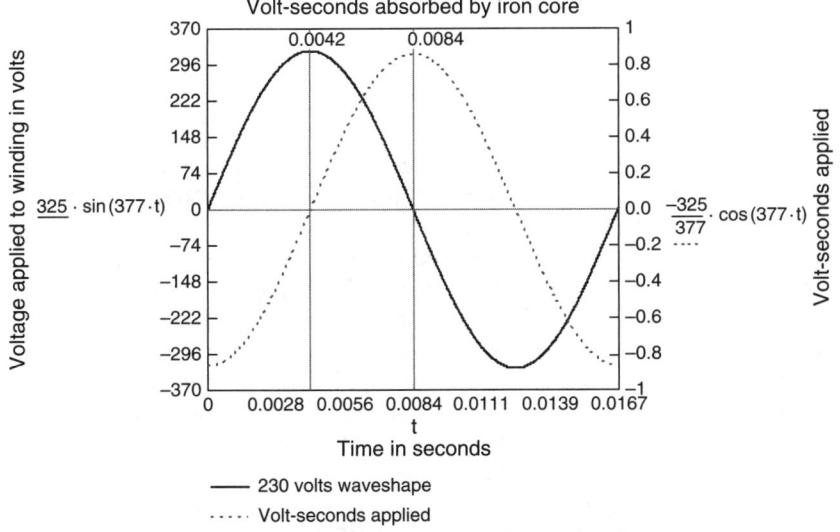

FIGURE 1-11 A 60-Hz voltage applied to winding and volt-seconds absorbed by the iron core.

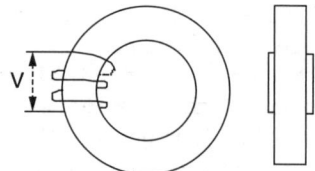

FIGURE 1-12 Toroidal iron core with one winding.

Integrating Eq. (1.9), we obtain the following

$$\int V_m \times \sin(\omega \times t)\, dt = \frac{-V_m}{\omega} \times \cos(\omega \times t) \quad (1.10)$$

The volt-seconds absorbed by the iron core during the positive half cycle of the applied voltage are

$$\int_0^{\frac{T}{2}} V_m \times \sin(\omega \times t)\, dt = \frac{-V_m}{\omega}(\cos\pi - \cos 0)$$

$$= \frac{-V_m}{\omega}(-1-1) = \frac{2 \times V_m}{\omega} = \frac{2\sqrt{2} \times V_{rms}}{\omega} \quad (1.11)$$

For 50 cycles: $\quad \dfrac{2\sqrt{2} \times V_{rms}}{314} = 0.0090 \times V_{rms} \quad$ half-cycle volt-seconds

For 60 cycles: $\quad \dfrac{2\sqrt{2} \times V_{rms}}{377} = 0.0075 \times V_{rms}$

For 400 cycles: $\quad \dfrac{2\sqrt{2} \times V_{rms}}{2513} = 0.001126 \times V_{rms}$

Keep in mind that in some applications the iron core does not need to absorb the volt-seconds of the entire half cycle. Also it is obvious that the 50-cycle sinusoidal wave contains more volt-seconds than the 60-cycle wave, and therefore the core cross section must be larger. This results in a heavier core. Consequently generators, motors, and transformers designed for 60 cycles, in general, cannot be used in 50-cycle power systems. However, devices designed for 50 cycles, in general, could be used in 60-cycle power systems. Incidentally, this is also one of the reasons for using 400 hertz (Hz) in aircraft applications, because magnetic devices are much lighter. Besides, breakers in a 400-Hz power system can react more quickly to a short circuit condition because the half cycle is only 1.25 ms long compared to 8.33 ms for the 60-Hz power system.

Each positive-going half cycle of the volt-seconds curve (the dotted line in Fig. 1-11) changes the iron core flux density B from a minimum to a maximum, which in many practical applications is from −16 to +16 kilogauss (kG). During the negative-going half cycle of the volt-seconds curve, the flux density changes from +16 to −16 kG. However, this down path of B does not retrace the up path of B, but rather returns following a curve above the up path, and therefore generates a loop called the *hysteresis loop*. The applied voltage generates 60 loops per second, and they are an important component of the iron losses. Furthermore, to keep the losses and the noise at a low level, it is convenient to use low flux density. However, this requirement must be weighed against the increase in size of the iron core. Using the induction law, it is possible to determine the flux density change produced in the iron core by a given amount of volt-seconds. Better still, a practical ΔB can be established, and the required NA_{Fe} can then be calculated.

Electromagnetic induction law:

$$e = N \times \left(\frac{d}{dt}\phi\right) \times 10^{-8} \text{ volts} \quad (1.12)$$

Integrating we obtain:

$$\int e\,dt = N \times \Delta\phi \times 10^{-8} = N \times A_{Fe} \times \Delta B \times 10^{-8} \text{ volt-second} \quad (1.13)$$

$$\Delta\phi = A_{Fe} \times \Delta B$$

If ΔB is expressed in kilogauss, the volt-seconds expression becomes:

$$\int e\,dt = N \times A_{Fe} \times \Delta B \times 10^{-5} \text{ volt-second} \quad (1.14)$$

where A_{Fe} = net cross sectional area of magnetic circuit, cm²
 ΔB = magnetic flux density change, kG
 N = number of turns of winding

Equation (1.14) is good for any wave shape and frequency. The volt-seconds absorbed by the iron core during a half cycle of a sinusoidal wave shape are:

$$\int_0^{\frac{T}{2}} V_m \times \sin(\omega \times t)\,dt = \frac{-V_m}{\omega}(\cos(\pi) - \cos(0))$$

$$= \frac{-V_m}{\omega}(-1-1) = \frac{2 \times V_m}{\omega} = \frac{2\sqrt{2} \times V_{rms}}{\omega} \quad (1.15)$$

Equating Eqs. (1.15) and (1.14) we obtain:

$$\frac{2 \times \sqrt{2} \times V_{rms}}{2 \times \pi \times f} = N \times A_{Fe} \times \Delta B \times 10^{-5}$$

$$N \times A_{Fe} = \frac{2\sqrt{2} \times V_{rms} \times 10^5}{2\pi \times f \times \Delta B} \text{ turns-cm}^2 \tag{1.16}$$

Applying Eq. (1.16) to a 60-Hz sinusoidal wave, we get

$$N \times A_{Fe} = \frac{2 \times \sqrt{2} \times 10^5}{377} \times \frac{V_{rms}}{\Delta B} = 750 \times \frac{V_{rms}}{\Delta B} \text{ turns-cm}^2 \tag{1.17}$$

Assuming $\Delta B = 32$ kilogauss and $V_{rms} = 230$ volts:

$$N \times A_{Fe} = 750 \frac{230}{32} = 539 \text{ turns-cm}^2 \tag{1.18}$$

Applying Eq. (1.16) to a 50-Hz sinusoidal wave, we get:

$$N \times A_{Fe} = \frac{2 \times \sqrt{2} \times 10^5}{314} \times \frac{V_{rms}}{\Delta B} = 901 \times \frac{V_{rms}}{\Delta B} \text{ turns-cm}^2 \tag{1.19}$$

$$N \times A_{Fe} = 901 \times \frac{230}{32} = 6476 \text{ turns-cm}^2$$

Applying Eq. (1.16) to a 400-Hz sinusoidal wave, we get

$$N \times A_{Fe} = \frac{2 \times \sqrt{2} \times 10^5}{2513} \times \frac{V_{rms}}{\Delta B} = 113 \times \frac{V_{rms}}{\Delta B} \text{ turns-cm}^2 \tag{1.20}$$

$$N \times A_{Fe} = 113 \times \frac{230}{32} = 812 \text{ turns-cm}^2$$

$$\frac{812}{5391} = 0.151$$

The $N \times A_{Fe}$ of the 400-Hz device is only 15 percent of the 60-Hz device.

CHAPTER 2

Power Systems Stability

The purpose of this chapter is to provide the theoretical background behind the formulas and procedures used in Chaps. 3 and 4.

2.1 Introduction

Synchronous generators are by far the most used type of generator in the power generation industry. They generate real and reactive power, and their output voltage is easy to control by changing the excitation or field current. Furthermore, in large power systems containing many generators, they rotate synchronously. There are two types of synchronous generator round-rotor and salient pole. They consist of a stator, on which the three-phase armature windings are wound 120° apart from one another, and a rotor, on which the DC field winding is wound. When the turbine rotates the rotor, the rotating magnetic flux induces sinusoidal voltages in the stator windings. The magnitudes of the voltages induced in the stator windings (which are 120° out of time phase from one another) depend on the magnitude of the field DC current and their frequencies depend on the rotor angular velocity. Round rotors are used in high-speed machines usually driven by high-speed steam or gas turbines (combustion turbines). This is so because round-rotor generators can better withstand the high centrifugal forces associated with turbine-driven high-speed machines; also their rotors are smaller in diameter and easier to dynamically balance. Salient pole generators are used for lower-speed applications, such as generators driven by water turbines. This type of machine needs several pairs of magnetic poles to generate power of 60 hertz (Hz). Fortunately, at the low speed at which they operate, the centrifugal forces are lower than those experienced by turbine-driven generators. Therefore, the rotor diameter could be larger and salient poles can be used. In round-rotor generators, the air gap between the stator and rotor is uniform with constant reluctance. However, in salient pole machines, the air gap varies along the generator circumference with the smaller gap along the direct axis of the rotor field winding and the larger gap along the neutral axis of the field winding, which is commonly called the *quadrature axis*. Therefore, in salient pole generators the reluctance of the air gap is not constant. To facilitate the mathematical analysis of salient pole generators, the magnetomotive force (mmf) is expressed in terms of its components along the direct and quadrature axes. The electromotive force (emf) produced by the changing flux linkage of each component of the acting mmf is considered as generated in two separate magnetic circuits, each with constant but different values of reluctance. This approach yields a salient pole generator model that consists of two parts, one for the direct axis and the other for the quadrature axis.

Electrical engineers and engineering professors alike are in the neglecting and assuming business. So when faced with a difficult problem—very complex and hard to clearly express in mathematical language, or the data required is hard to get or nonexistent—we simplify the problem by neglecting things that we believe will not change the outcome very much. Actually, we keep neglecting until we can arrive at a "solution." That is the approach we take to analyze the transient stability of a multimachine power system. We use the classical (and very simple) model to represent any generator connected to the network. In a power system operating in stable condition, the angular positions of the synchronous machines' rotors remain constant relative to one another when no disturbance occurs. *Small-disturbance stability* occurs when a power system operating in a steady-state stable condition, following some small disturbance, returns to the same steady-state operating condition or very close to it. Furthermore, the power system is *transiently stable* if, following a large or sudden disturbance, it reaches a different but acceptable steady-state operating condition.

2.2 Classical Model

The classical synchronous generator model is shown in Fig. 2-1. This simple model is by far the most used in stability studies, although it is not the best machine representation. In fact, the classical model is used even in the case of generators with salient pole rotors. In the classical model, all phasors, such as generated voltage and current, are usually expressed with reference to the generator terminal voltage, and the resistive portions of all the impedances are neglected. Also, in the classical model, the generator reactance is easily combined by simple addition with the network reactance, such as the reactance of transformers and transmission lines. In generators with salient pole rotors, the direct-axis transient reactance is different from the quadrature transient reactance. But the fact that the reactance of the transmission line connecting the generator with the load busses is usually larger than the

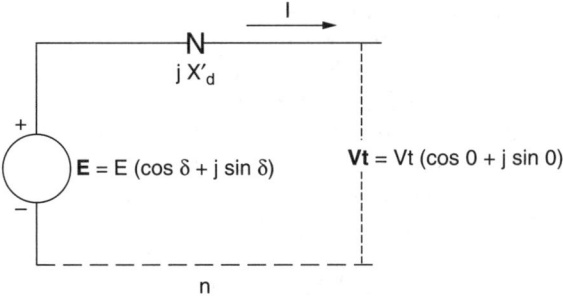

FIGURE 2-1 Classical generator model.

transient values of the generator's direct and quadrature reactance validates the use of the classical model to represent machines with salient poles and nonuniform flux linkages. Sometimes, in stability studies, salient pole generator models consist of two classical models, one for the direct-axis transient reactance and the other for the quadrature transient reactance. Although the two magnetic circuits could be considered separate from each other, the electric circuits of both models are interconnected in parallel and can be resolved into a single one. In reality, two diagrams complicate the computations.

When the load connected to the network suddenly changes or a fault occurs, the main field flux linkages and the generated emf are assumed constant for a short time of around 1 second, although the flux is decaying as determined by the time constant of the field winding circuit. Actually, the armature reaction (Lenz' law) plus the response of the excitation system trying to control the network disturbance tends to maintain the flux and the generated emf constants for at least the assumed 1 second. This period of time is large enough for stability studies concerned with the rotor's first swing. In these studies it is valid to consider the generated emf constant. The problem with the number of swings is that the classical model and the network representation itself are only good for linear parameters. But when a fault occurs, the magnitude of the current increases approximately from 10 to 16 times the normal generator rating, and this high value of the armature current could saturate the magnetic circuit. In addition, harmonics are introduced into the power system with the consequent voltage distortions. In particular, the even harmonics produce a DC offset that saturates all magnetic circuits, such as transformers' cores, reactors, and the generator's stator. So the system becomes very quickly nonlinear, and therefore the classic model and associated computation procedures are invalid after the first rotor swing or after 1 second if you are optimistic.

The classical model of Fig. 2-1 shows the emf **E** in series (behind) with the generator transient reactance X'_d. This simple model with a single electrical diagram to represent the generator, instead of two separate ones for the case of salient pole rotors, is based on the assumption that $X'_d = X'_q$ or that the direct-axis transient reactance is equal to the quadrature transient reactance even during transient conditions. However, the results obtained using the single diagram are acceptable because usually the larger value of the reactance of the electrical network, the step-up transformer included, makes any difference between X'_d and X'_q insignificant. The *generated voltage*, which is assumed constant for the duration of the first swing (or for 1 second approximately), is defined as the value of **E** during the transient. The power angle δ is the angle between **E** and **Vt** where **Vt** is designated as the phasor reference. In this manner each generator would have its own individual phasor reference. So to avoid having a power system with several phasor references, the neutral of the power system, to which all the generators are connected, is selected as the common reference for the entire power system, and it is assumed that this common reference rotates at synchronous angular speed.

2.3 Power Flow from Generator to Motor

The power flow in the network illustrated in Fig. 2-2 is given in terms of the generalized circuit constants A, B, C, D, and they include the synchronous impedance of both machines as well as the impedance of the circuit connecting them. Also the nominal π interconnecting the machines is assumed linear and passive and has two pairs of terminals. The reader should keep in mind that the linearity of the network components is only an ideal concept to facilitate the analysis and computations. Rigorously speaking, each shunt admittance of the nominal π representation should be shunted by a pure resistance to account for the leakage current between conductors, due to air humidity or imperfect insulation.

In addition, the always necessary transformers are nonlinear devices owing to iron core saturation produced by even harmonics, or the DC offset during short circuit conditions. However, the fact

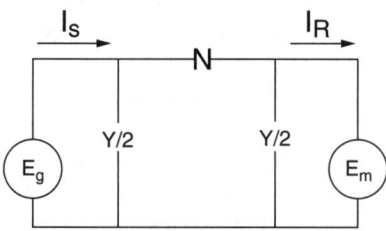

Figure 2-2 Network equivalent symmetrical π.

that the analysis is for steady-state conditions makes the assumption that all the components of the network are linear acceptable.

$$E_g = \left(I_R + \frac{Y}{2} \cdot E_m\right) \cdot Z + E_m$$

$$E_g = \left(\frac{Z \cdot Y}{2} + 1\right) \cdot E_m + Z \cdot I_R \tag{2.1}$$

$$E_g = A \cdot E_m + B \cdot I_R \tag{2.2}$$

$$I_S = \frac{Y}{2} \cdot E_g + \frac{Y}{2} \cdot E_m + I_R = \left[\left(\frac{Z \cdot Y}{2} + 1\right) \cdot E_m + Z \cdot I_R\right] \cdot \frac{Y}{2} + E_m \cdot \frac{Y}{2} + I_R$$

$$I_S = Y\left(\frac{Z \cdot Y}{4} + 1\right) \cdot E_m + \left(\frac{Z \cdot Y}{2} + 1\right) \cdot I_R \tag{2.3}$$

$$I_S = C \cdot E_m + D \cdot I_R \tag{2.4}$$

The generalized network constants for the nominal symmetrical π network are

$$A = \frac{Z \cdot Y}{2} + 1 \qquad B = Z \qquad C = Y\left(\frac{Z \cdot Y}{4} + 1\right) \qquad D = \frac{Z \cdot Y}{2} + 1 \tag{2.5}$$

These constants are applicable to bilateral networks having two pairs of terminals and formed by linear and passive components. Both A and D are unitless, B is in ohms, and C is in mho. And they are related by the following equation:

$$A \cdot D - B \cdot C = 1 \tag{2.6}$$

The receiving-end current is obtained from Eq. (2.2)

$$I_R = \frac{E_g - A \cdot E_m}{B} \tag{2.7}$$

where the following notation is assigned:

$$A = |A| \cdot e^{j \cdot \alpha} \qquad B = |B| \cdot e^{j \cdot \beta} \qquad E_g = |E_g| \cdot e^{j \cdot \delta} \qquad E_m = |E_m| \cdot e^{j \cdot 0} \quad \text{reference}$$

$$I_R = \frac{|E_g| \cdot e^{j \cdot \delta} - |A| \cdot |E_m| \cdot e^{j \cdot \alpha}}{|B| \cdot e^{j \cdot \beta}} = \left|\frac{E_g}{B}\right| \cdot e^{j(\delta - \beta)} - \left|\frac{A \cdot E_m}{B}\right| \cdot e^{j(\alpha - \beta)} \tag{2.8}$$

$$\overline{I_R} = \left|\frac{E_g}{B}\right| \cdot e^{j(\beta - \delta)} - \left|\frac{A \cdot E_m}{B}\right| \cdot e^{j(\beta - \alpha)} \qquad \text{conjugate of } I_R$$

The complex power at the receiving end is $E_m \cdot \overline{I_R}$

$$P_m + j \cdot Q_m = \frac{|E_g| \cdot |E_m|}{|B|} \cdot e^{j(\beta-\delta)} - \frac{|A| \cdot (|E_m|)^2}{|B|} \cdot e^{j(\beta-\alpha)} \qquad (2.9)$$

Angles α and β are parameters of the connecting transmission lines and their values depend on series and shunt line impedances. Therefore, in Eq. (2.9) they are not variable.

Formulas: $\quad e^{j\theta} = \cos\theta + j \cdot \sin\theta \qquad e^{-j\theta} = \cos\theta - j \cdot \sin\theta$

$$P_m = \frac{|E_g| \cdot |E_m|}{|B|} \cos(\beta - \delta) - \frac{|A| \cdot (|E_m|)^2}{|B|} \cdot \cos(\beta - \alpha) \quad \text{real power} \quad (2.10)$$

$$Q_m = \frac{|E_g| \cdot |E_m|}{|B|} \sin(\beta - \delta) - \frac{|A| \cdot (|E_m|)^2}{|B|} \cdot \sin(\beta - \alpha) \quad \text{reactive power} \quad (2.11)$$

Equations (2.9) to (2.11) provide the complex, real, and reactive, power received by the motor. In stability studies the absolute values of E_g and E_m are considered constant, which is a fair assumption because the output voltage of the generator is regulated. Therefore, the only variable is δ, the power or torque angle. This, for a given generator, is the angle between the rotor axis and the stator magnetic field axis. When the load changes suddenly, the generator rotor will accelerate or decelerate with respect to the rotating stator field, changing the power angle. In analytical studies with synchronous generator models, the power angle is the angle between two phasors: the generated emf and the generator's terminal voltage (see Fig. 2-1).

As the load on the motor changes, the power transferred from the generator to the motor changes. To this change in power corresponds a change in the relative angular position of the machine's rotors, or a change in δ. Equation (2.10) shows that the maximum real power that could be transferred to the motor occurs when $\delta = \beta$. This is expressed symbolically by Eq. (2.12). If the power-transferring network contains any resistance or shunt admittance, as the nominal π does, then the maximum power that the motor could receive from the generator is less than the maximum power that the generator could deliver to the network. If the motor load changes, the power angle between the generator and the motor changes, then when the motor load decreases, δ decreases and when the load increases, δ increases. However, if the power required by the load is greater than the maximum power that could be transferred from the generator to the motor, which happens when $\delta = \beta$, then the motor will slow down as the additional power required by the load comes from the stored rotational energy in the motor-load combination. As the motor slows down, it reaches a point at which δ becomes greater than β (is no longer equal to β), causing a decrease in the generated power and a further increase in δ and eventually falling out of synchronism with the generator. So Eq. (2.12) also provides the steady-state stability limit of the system shown in Fig. 2-2, which is defined as the maximum electric power that can be transferred to the receiving end without loss of synchronism. To summarize, Eq. (2.12) provides the maximum real power that could be delivered to the motor and the steady-state stability limit; both occur when $\delta = \beta$:

$$P_{m,max} = \frac{|E_g| \cdot |E_m|}{|B|} - \frac{|A| \cdot (|E_m|)^2}{|B|} \cdot \cos(\beta - \alpha) \qquad (2.12)$$

Neglecting the resistances and shunt admittances makes $\beta = \pi/2$ and $Y = 0$; we then arrive at the system portrayed in Fig. 2-3, where X includes the per-unit values of the line reactance and of the transient reactances of the generator and motor. We selected transient reactance instead of the steady-state reactance because the machine's rotor is continuously changing position with respect to the flux and magnetomotive force produced by the stator current. Applying Eq. (2.5), we obtain the specific values of the generalized circuit constant for this network.

Shunt capacitive reactance:

$$X_C = \frac{1}{\omega \cdot C} \qquad \text{If} \quad C = 0 \quad X_C = \infty \quad Y_C = 0$$

$$A = \frac{Z \cdot Y}{2} + 1 \qquad B = Z \qquad C = Y\left(\frac{Z \cdot Y}{4} + 1\right) \qquad D = \frac{Z \cdot Y}{2} + 1$$

See Eq. (2.5). The assumptions made ($Z = X, Y = 0$) and the fact that 1 is a real number, convert the expression of the generalized network constants for the nominal π to

$$A = 1 \cdot e^{j\alpha} \qquad B = |X| \cdot e^{j\beta} \qquad C = 0 \qquad D = 1 \cdot e^{j\alpha}$$

where $\qquad \alpha := 0 \qquad \beta := \frac{\pi}{2} \qquad \cos(\beta - \alpha) = 0$

$$A = D = 1 \qquad B = |X| \cdot e^{j\pi/2} \qquad C = 0 \qquad (2.13)$$

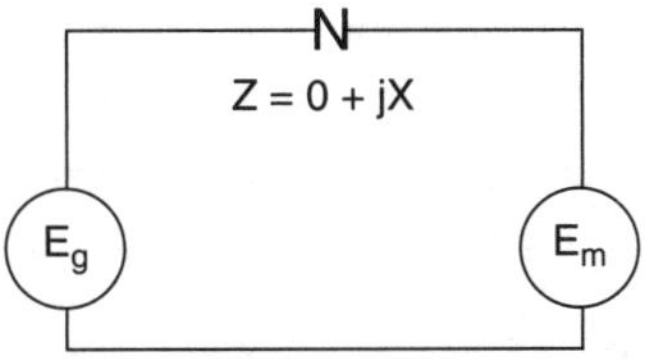

FIGURE 2-3 Network impedance diagram.

Substituting the values shown in Eq. (2.13) in Eqs. (2.10) and (2.11), we obtain

$$P_m = \frac{|E_g|\cdot|E_m|}{|X|}\cos\left(\frac{\pi}{2}-\delta\right) - \frac{|A|\cdot(|E_m|)^2}{|X|}\cdot\cos\frac{\pi}{2}$$

$$P_m = \frac{|E_g|\cdot|E_m|}{|X|}\sin\delta \quad \text{real power} \tag{2.14}$$

$$P_{m,max} = \frac{|E_g|\cdot|E_m|}{|X|} \quad \text{maximum real power and steady-state stability limit} \tag{2.15}$$

$$P_m = P_{m,max}\cdot\sin\delta$$

$$Q_m = \frac{|E_g|\cdot|E_m|}{|X|}\sin\left(\frac{\pi}{2}-\delta\right) - \frac{(|E_m|)^2}{|x|}\cdot\sin\frac{\pi}{2} \tag{2.16}$$

$$Q_m = \frac{|E_g|\cdot|E_m|}{|X|}\cos\delta - \frac{(|E_m|)^2}{|x|} \quad \text{reactive power} \tag{2.17}$$

When the motor is receiving the maximum real power, the reactive power it receives is found by substituting $\delta = \beta = \pi/2$ in Eq. (2.17). Symbolically,

$$Q_m = -\frac{(|E_m|)^2}{|x|} \tag{2.18}$$

That is an acceptable operating point. The minus sign indicates that when the motor receives maximum real power, the network (motor included) sends back the indicated reactive power.

From Eq. (2.15) the following conclusions regarding the stability limit are obvious:

1. Increasing $|E_g|$, $|E_m|$, or both, increases the maximum electric power that could be transfered and consequently the steady-state stability limit. This could be accomplished by increasing the excitation of the synchronous machines. However, connected equipment and insulation ratings limit the amount that the voltage could be increased. Furthermore, Eq. (2.14) shows that if the motor's load remains constant when the absolute value of the machine's internal voltages are increased, then the torque angle δ will decrease, and that is a welcome change.

2. Reducing $|X|$ increases the maximum power that could be safely transmitted as well as the steady-state stability limit. This must be carefully considered when selecting the network's equipment (transformers, generator, motor, and conductor's size and separation). However, if the equipment is already

installed, then the simplest way to reduce the overall reactance of the network is to install a parallel transmission line capable of carrying the entire load. A parallel line will increase the reliability of the network since one line could carry the entire load in case of a fault in the other. The additional line will also enhance the stability of the system because the more power the system transmits during fault conditions the more stable it will be. Another way of reducing the line voltage drop and increasing the stability limit is to connect capacitors in series with the line to decrease the total reactance.

2.4 Steady-State Stability

When a power system operating in steady-state condition remains in synchronism following a small disturbance and returns to the same (voltage magnitude and power angle) steady-state operating condition it is classified as *steady-state stable*.

The steady-state stability of a power system is determined by using a linear system model and assuming the following:

- Voltage regulators and other automatic controls are not active.
- Power changes are small.
- Voltage angle changes are small.

Actually, very seldom are the automatic controls off. Besides, the small disturbances are considered unimportant, except for the manufacturers of automatic control devices.

The swing equation for a small disturbance of the power angle is

$$\delta = \delta_0 + \Delta\delta$$

where $\quad \Delta\delta \ll \delta \quad \cos\Delta\delta \approx 1 \quad \sin\Delta\delta \approx \Delta\delta$

$$\frac{H}{\pi \cdot f_0} \cdot \left[\frac{d^2}{dt^2}(\delta_0 + \Delta\delta)\right] = P_{me} - P_{max} \cdot \sin(\delta_0 + \Delta\delta)$$

where P_{me} = mechanical power and P_{max} = maximum electric power. The above swing equation could be converted to

$$\frac{H}{\pi \cdot f_0} \cdot \frac{d^2}{dt^2}\Delta\delta + P_{max} \cdot \cos\delta_0 \cdot \Delta\delta = 0$$

The steady-state term is equal to zero

$$\frac{H}{\pi \cdot f_0} \cdot \frac{d^2}{dt^2} \cdot \Delta\delta + P \cdot \Delta\delta = 0 \qquad (2.19)$$

where $P = P_{max} \cdot \cos\delta_0$.

2.5 Brief Summary of Rotational Dynamics

Stability studies of rotating rigid bodies that are symmetrical about a fixed rotational axis requires the application of the mechanical formulas listed in Table 2-1, which also includes the corresponding formulas for similar concepts in linear motion.

$$\text{Linear motion: } F = m\frac{d^2}{dt^2}S \quad \text{mass times linear acceleration}$$

$$\text{Rotational motion: } T = I\frac{d^2}{dt^2}\theta \quad \text{moment of inertia times angular acceleration} \quad (2.20)$$

Applying Eqs. (2.20) to synchronous generators, we obtain

$$I\frac{d^2}{dt^2}\theta = T_a = T_s - T_e \quad (2.21)$$

where I = total moment of inertia of rotational masses in generator and prime mover
 θ = angular position of rotor with respect to fixed reference in mechanical radians
 t = time, in second
 T_a = net accelerating torque
 T_s = shaft torque delivered by prime mover minus torque representing rotational losses
 T_e = net electromagnetic torque produced by generator

A synchronous generator delivers power to the electrical system, and we define the shaft torque T_s and the electromagnetic torque T_e as positive. Starting up the generator requires increasing the shaft torque to accelerate the generator rotor in the positive θ direction of rotation, generating a positive electromagnetic torque which always opposes the shaft torque and an electromotive force of the right polarity.

When both torques are equal, the acceleration torque T_a is zero, there is no positive or negative acceleration, and the rotor speed is then the synchronous speed. In this case, both the electromagnetic torque and the generated emf are constant.

A synchronous motor receives power from the electrical system, and consequently T_s and T_e are inverted in sign. In this case T_e represents the driving torque produced by the power delivered to the motor, and T_s represents the opposing countertorque of the mechanical load and rotational losses. During the starting period T_e must be larger than T_s, but they are equal when the motor rotor reaches synchronous speed.

The torque T, the angular velocity ω, and the angular acceleration α are all vectors acting along the same axis. Therefore we can consider only their magnitudes and treat them algebraically.

The value of M at synchronous speed is called the *inertial constant*, which unfortunately is also the name assigned to H. In dealing with

Quantity	Symbol	Formula	Unit	Quantity	Symbol	Formula	Unit
Length	S		m	Angular displacement	θ	$\theta = \dfrac{S}{r}$	radians
Mass	m	$m = \dfrac{W}{g}$	kgm	Moment of Inertia	I	$I = \int r^2 \, dm$	$\dfrac{\text{joule} \cdot s^2}{\text{rad}^2}$
Velocity	v	$v = \dfrac{d}{dt} S$	$\dfrac{m}{s}$	Angular velocity	ω	$\omega = \dfrac{d}{dt}\theta$	$\dfrac{\text{rad}}{s}$
Acceleration	a	$a = \dfrac{d}{dt} v$	$\dfrac{m}{s^2}$	Angular acceleration	α	$\alpha = \dfrac{d}{dt}\omega$	$\dfrac{\text{rad}}{s^2}$
Force	F	$F = m \cdot a$	N	Torque	T	$T = I \cdot \alpha$	$\dfrac{J}{\text{rad}}$
Momentum	p	$p = mv$	$N \cdot s$	Angular momentum	M	$M = I \cdot \omega$	$\dfrac{J \cdot s}{\text{rad}}$
Work	W	$W = \int F \, dS$	J	Work	W	$W = \int T \, d\theta$	J
Trans kinetic energy	KE	$KE = \dfrac{m \cdot v^2}{2}$	J	Rot kinetic energy	KE	$\dfrac{I \cdot \omega^2}{2}$	J
Power	P	$P = F \cdot v$	w	Power	P	$P = T \cdot \omega$	w

TABLE 2-1 Mechanical Formulas Used to Analyze the Transient Stability of Electrical Networks

generators, it is better to express the stored rotational kinetic energy in megajoules- or megawatt-seconds.

The angular momentum M is:

$$M = I \cdot \omega \qquad \frac{\text{megajoules} \times \text{second}}{\text{radian}} \quad \text{or} \quad \frac{\text{megajoules} \times \text{second}}{\text{degree}}$$

Stored rotational energy:

$$\frac{I \cdot \omega^2}{2} = \frac{M \cdot \omega}{2} \quad \text{megajoules- or megawatt-second}$$

The inertial constant H is the stored rotational energy of the generator at synchronous speed in megajoules per unit of generator rating in megavolt-amperes (MVA).

H = stored rotational energy, megajoules/generator MVA rating

1 megajoule (MJ) = 1 megawatt-second (MW·s),

so H could also be expressed as

H = stored rotational energy, megawatt-second/generator MVA rating

Symbolically,

$$H = \frac{M \cdot \omega_s}{2G} \qquad \text{G is generator rating, MVA}$$

GH = stored rotational energy at synchronous speed, in megajoules- or megawatt-seconds.

$$GH = \frac{M \cdot \omega_s}{2} \quad \text{megawatt-seconds} \tag{2.22}$$

For generators with one pair of poles and a frequency of 60 cycles per second (cycles/s), and expressing the angles in radians (rad), we have

$$f := 60 \, \text{cycle/sec} \qquad \omega_s := 2 \cdot \pi \cdot f \qquad \omega_s = 377 \, \text{rad/sec}$$

$$\frac{\omega_s}{2} = \pi \cdot f = 188.5 \, \text{rad/sec}$$

$$G \cdot H = M \cdot \pi \cdot f = 188.5 \quad \text{megajoules- or megawatt-second} \tag{2.23}$$

$$M = \frac{G \cdot H}{\pi \cdot f} \qquad \frac{\text{megawatt} \times \text{second}^2}{\text{rad}} \tag{2.24}$$

A gain for generators with one pair of poles and frequency of 60 cycles per second, expressing the angles in degrees, is

$$f := 60 \, \frac{\text{cycles}}{\text{sec}} \qquad \omega_s := 360 \cdot f = 21{,}600 \, \frac{\text{degree}}{\text{sec}}$$

$$\frac{\omega_s}{2} = 180 \cdot f = 10{,}800 \, \frac{\text{degree}}{\text{sec}}$$

Using Eq. (2.22) with electrical degrees, we obtain Eqs. (2.25) and (2.26).

$$G \cdot H = M \cdot 180 \cdot f \qquad M = \frac{G \cdot H}{180 f} \qquad \frac{\text{megawatt} \times \text{second}^2}{\text{degree}} \qquad (2.25)$$

$$M = 0.000093 \, G \cdot H \qquad \frac{\text{megawatt} \times \text{second}^2}{\text{degree}} \qquad (2.26)$$

The inertial constant H of a synchronous machine has a small range of values for each type of machine independent of its MVA and angular speed ratings. In fact, for all types of machines it ranges from 1 to 10 megawatt-second per MVA of the machine rating. In addition, H can be calculated provided that the weight and radius of gyration of all the rotating parts of the generator and prime mover are known. The manufacturers of synchronous generators usually provide enough data to compute H.

$$KE = \frac{I \cdot \omega_m^2}{2} \qquad \text{rotational kinetic energy where } \omega_m \text{ is mechanical angular speed and I is moment of inertia}$$

For 3600 rpm machines we have

Data: 3600 rpm $\qquad g = 32.17 \text{ ft/sec}^2$

$$\omega_m = 2 \cdot \pi \cdot \frac{3600}{60} = 120 \cdot \pi = 377 \text{ rad/sec}$$

$$I = \frac{W \cdot R^2}{32.17} \qquad \text{W in pounds, R in feet}$$

$$KE = \frac{1}{2} \cdot \frac{W \cdot R^2}{32.17} \cdot 377^2 \qquad \text{foot-pounds (ft} \cdot \text{lb)}$$

To convert the rotational energy from foot-pounds to megajoules, multiply by 1.3564×10^{-6}.

$$KE = \left(\frac{1}{2} \cdot \frac{W \cdot R^2}{32.17} \cdot 377^2 \right)(1.3564 \times 10^{-6}) \text{ megajoules}$$

$$\left(\frac{1}{2} \cdot \frac{W \cdot R^2}{32 \cdot 17} \cdot 377^2 \right)(1.3564 \times 10^{-6}) \rightarrow 3 \times 10^{-3} \cdot R^2 \cdot W \qquad \text{megajoules}$$

$$H = \frac{3 \times 10^{-3} \cdot R^2 \cdot W}{G} \qquad \frac{\text{megajoules}}{\text{MVA}} \qquad 3600 \text{ rpm case} \qquad (2.27)$$

For 1800 rpm machines:

$$\omega_m = 2\cdot\pi\cdot\frac{1800}{60} = 60\cdot\pi = 188.5\,\frac{\text{rad}}{\text{sec}} \qquad \text{everything else the same}$$

$$\left(\frac{1}{2}\cdot\frac{W\cdot R^2}{32.17}\cdot 188.5^2\right)(1.3564\times 10^{-6}) \rightarrow 7.49\times 10^{-4}\cdot R^2\cdot W \quad \text{megajoules}$$

$$H = \frac{7.49\times 10^{-4}\cdot R^2\cdot W}{G} \qquad \frac{\text{megajoules}}{\text{MVA}} \qquad 1800\,rpm\,case \qquad (2.28)$$

For 900 rpm machines:

$$\omega_m = 2\cdot\pi\cdot\frac{900}{60} = 30\cdot\pi = 94.25\,\frac{\text{rad}}{\text{sec}} \qquad \text{everything else the same}$$

$$\left(\frac{1}{2}\cdot\frac{W\cdot R^2}{32.17}\cdot 94.25^2\right)(1.3564\times 10^{-6}) \rightarrow 1.87\times 10^{-4}\cdot R^2\cdot W \quad \text{megajoules}$$

$$H = \frac{1.87\times 10^{-4}\cdot R^2\cdot W}{G} \qquad \frac{\text{megajoules}}{\text{MVA}} \qquad 900\,rpm\,case \qquad (2.29)$$

2.6 The Swing Equation

In Sec. 2.5, the angular position of the rotor θ is measured with respect to a fixed reference on the stator, and therefore it grows continuously with time. The important thing, however, is the angular position of the rotor with respect to a reference axis that rotates at synchronous speed. Symbolically,

$$\theta = \omega_s\cdot t + \delta \qquad (2.30)$$

where θ = angular position of rotor with respect to fixed reference, mechanical radians
ω_s = generator synchronous speed, mechanical radians per second
δ = rotor angular displacement from synchronous rotating reference axis, mechanical radian

Actually Eq. (2.30) is also valid if the angles are expressed in mechanical degrees. Equation (2.31) is obtained by assuming ω_s constant, which is a fair assumption, and writing the derivative with respect to t:

$$\frac{d}{dt}\theta = \omega_s + \frac{d}{dt}\delta \qquad (2.31)$$

Equation (2.31) gives the rotor angular velocity, and it shows that it is constant and equal to the reference synchronous speed ω_s only when

$$\frac{d}{dt}\delta = 0$$

So $(d/dt)\delta$ is the deviation from synchronism. When $(d/dt)\delta$ is constant, the generator is running at a constant angular speed equal to the synchronous angular speed, ω_s, plus $(d/dt)\delta$.

$$\frac{d^2}{dt^2}\theta = \frac{d^2}{dt^2}\delta \tag{2.32}$$

Equation (2.32) provides the rotor angular acceleration, and it shows that is equal to the speed of the deviation. Furthermore it is valid in any consistent set of angular units, mechanical or electrical. Substituting in Eq. (2.21), we obtain

$$I \cdot \frac{d^2}{dt^2}\delta = T_a = T_s - T_e \tag{2.33}$$

$$\alpha = \frac{d^2}{dt^2}\delta \quad \text{acceleration} \quad \alpha \cdot I = T_a$$

From Table 2-1 we know that $M = \omega I$ and $P = T\omega$. Multiplying both sides of Eq. (2.33) by ω, we obtain

$$\omega \cdot I \cdot \frac{d^2}{dt^2}\delta = P_a = P_s - P_e \tag{2.34}$$

$$M \cdot \frac{d^2}{dt^2}\delta = P_a = P_s - P_e \quad \textit{swing equation} \tag{2.35}$$

where M = rotor angular momentum at synchronous speed
 δ = torque angle or rotor angular displacement from synchronous rotating reference
 P_a = accelerating power
 P_s = net mechanical shaft power
 P_e = net electric power generated

Another way of presenting the swing equation is derived below: From Eq. (2.22) we obtain

$$M = \frac{2 \cdot H}{\omega_s} \cdot G$$

where G is the generator rating in MVA. Substituting in Eq. (2.35), we obtain

$$\frac{2 \cdot H}{\omega_s} \cdot \frac{d^2}{dt^2}\delta = \frac{P_a}{G} = \frac{P_s - P_e}{G} \quad \begin{array}{l}\delta \text{ and } \omega_s \text{ should be expressed in}\\ \text{consistent set of units: radians or degrees}\end{array}$$

$$\frac{2 \cdot H}{\omega_s} \cdot \frac{d^2}{dt^2}\delta = P_a = P_s - P_e \quad P_a, P_s, \text{ and } P_e \text{ are in per unit} \tag{2.36}$$

When δ is in electrical radians, Eq. (2.36) becomes

$$\frac{H}{\pi \cdot f} \cdot \frac{d^2}{dt^2} \delta = P_a = P_s - P_e \quad \text{per unit} \qquad (2.37)$$

When δ is in electrical degrees, Eq. (2.36) becomes

$$\frac{H}{180 \cdot f} \cdot \frac{d^2}{dt^2} \delta = P_a = P_s - P_e \quad \text{per unit} \qquad (2.38)$$

In the case of a generator and infinity bus system, the swing equation becomes Eq. (2.39) by substituting Eq. (2.14) for P_e in Eq. (2.35).

$$M \cdot \frac{d^2}{dt^2} \delta = P_s - \frac{|E_g| \cdot |E_m|}{|X|} \cdot \sin \delta \qquad (2.39)$$

According to Eq. (2.15), the maximum power that could be transferred without exceeding the stability limit is

$$P_{max} = \frac{|E_g| \cdot |E_m|}{|X|}$$

So Eq. (2.39) becomes

$$M \cdot \frac{d^2}{dt^2} \delta = P_s - P_{max} \cdot \sin \delta \qquad (2.40)$$

The swing equation is a second-order differential equation that could be solved by converting it to a system of two coupled first-order differential equations and applying any of the readily available differential equation numerical solvers. The author uses MathCad 14. The solution is given in matrix (table) format and shows the power angle δ increasing with time, a clear indication that the system could be unstable for the disturbance considered. In that case the disturbance (a fault or large, sudden load increase) must be cleared before the power angle reaches the critical value δ_c to ensure that the system will remain stable after the fault is cleared.

2.7 Synchronizing Power Coefficient

Let us consider the simplified network illustrated in Fig. 2-3, where E_g is the generated output voltage and E_m is the voltage at the infinity bus or the voltage delivered to a synchronous motor. The reactance shown is the sum of the machine's synchronous reactance and the transmission line reactance. All the resistances have been neglected. The real power delivered by the generator is given by Eq. (2.14). Let us say that the machine is operating in steady-state condition at a power angle δ_0. The electric power delivered by the generator in this initial condition is

$$P_{e0} = \frac{|E_g| \cdot |E_m|}{|X|} \cdot \sin \delta_0 \qquad \text{initial electric power transferred between generator and infinity bus} \qquad (2.41)$$

$$P_{e,max} = \frac{|E_g| \cdot |E_m|}{|X|} \qquad \text{maximum electric power that generator could deliver} \qquad (2.42)$$

$$P_{e0} = P_{e,max} \cdot \sin \delta_0 \qquad \text{initial value of electric power angle equation} \qquad (2.43)$$

The steady-state operating point of a generator must be such that it should not lose synchronism when a small change occurs in its electric power output. Let us consider that the shaft power remains constant when the output of electric power changes. Assuming incremental changes in the parameters defining the initial operating point, we have

$$\delta = \delta_0 + \delta_\Delta \qquad P_e = P_{e0} + P_{e\Delta} \qquad P_s = \text{constant}$$

where δ_0 = initial operating point power angle
δ_Δ = change in power angle
δ = new power angle
P_{e0} = electric output
$P_{e\Delta}$ = power initial
P_e = operating point

From Eq. (2.14) we get

$$P_{e0} + P_{e\Delta} = P_{e,max} \cdot \sin(\delta_0 + \delta_\Delta)$$
$$= P_{e,max} \cdot \cos \delta_0 \cdot \sin \delta_\Delta + P_{e,max} \cdot \cos \delta_\Delta \cdot \sin \delta_0$$

If δ_Δ is small enough, we can say that $\sin \delta_\Delta \approx \delta_\Delta$ and $\cos \delta_\Delta \approx 1$

$$P_{e0} + P_{e\Delta} = P_{e,max}(\delta_\Delta \cdot \cos \delta_0 + \sin \delta_0)$$
$$P_{e0} + P_{e\Delta} = P_{e,max} \cdot \cos \delta_0 \cdot \delta_\Delta + P_{e,max} \cdot \sin \delta_0 \qquad (2.44)$$

where $P_{e,max} \cdot \cos \delta_0$ is defined as the synchronizing power coefficient. Symbolically,

$$P_{sy} = P_{e,max} \cdot \cos \delta_0 \qquad \text{synchronizing power coefficient} \qquad (2.45)$$

From Eqs. 2.41 to 2.44 we obtain

$$P_{e0} + P_{e\Delta} = \delta_\Delta \cdot P_{sy} + P_{e0} \qquad P_{e\Delta} = P_{sy} \cdot \delta_\Delta \qquad (2.46)$$

The increment of the power angle times the synchronizing coefficient provides the increment in electric power delivered by the generator.

At the initial operating point when the power angle is δ_0, the generator is operating in steady-state condition, and therefore the shaft power is equal to the electric power generated. Symbolically,

$$P_s = P_{e0} = P_{e,max} \cdot \sin \delta_0 \quad (2.47)$$

Subtracting Eq. (2.44) from Eq. (2.47), we obtain

$$P_s - (P_{e0} + P_{e\Delta}) = -P_{e,max} \cdot \cos \delta_0 \cdot \delta_\Delta \quad (2.48)$$

Expressing the swing equation, Eq. (2.36), in terms of the new value of the variables, we obtain

$$\frac{2 \cdot H}{\omega_S} \cdot \frac{d^2}{dt^2}(\delta_0 + \delta_\Delta) = P_s - (P_{e0} + P_{e\Delta}) \quad (2.49)$$

Combining Eqs. (2.48) and (2.49) gives

$$\frac{2 \cdot H}{\omega_S} \cdot \frac{d^2}{dt^2}(\delta_0 + \delta_\Delta) + P_{e,max} \cdot \cos \delta_0 \cdot \delta_\Delta = 0 \quad (2.50)$$

The synchronizing power coefficient has been defined in Eq. (2.45) as $P_{sy} = P_{e,max} \cdot \cos \delta_0$. Since $P_{e,max} \cdot \cos \delta_0$ is the slope of the power angle curve at δ_0, see Eq. (2.43). Symbolically,

$$P_{sy} = \left(\frac{d}{d\delta} P_e\right)_{\delta=\delta_0} = P_{e,max} \cdot \cos \delta_0$$

Since δ_0 is a constant, Eq. (2.50) can be expressed as

$$\frac{d^2}{dt^2}\delta_\Delta + \frac{\omega_S}{2 \cdot H} \cdot P_{sy} \cdot \delta_\Delta = 0 \quad (2.51)$$

The system steady-state stability for small load changes is determined by Eq. (2.51), which is a linear second-order differential equation whose solution depends on the sign of P_{sy} or the sign of the slope of the power angle curve at the operating point.

If P_{sy} is positive, the increment of the power angle as a function of time $\delta_\Delta(t)$ is oscillatory about δ_0, like an undamped swinging pendulum. However, the ever-present resistance in transmission lines, transformers, and the damper windings of the generator, which have been all neglected, introduce decay in the amplitude of the oscillations that eventually will stop the oscillations of the power angle. On the other hand, if P_{sy} is negative, the power angle will increase without a limit after the occurrence of small change in the load. Therefore the system would be steady-state stable only if the slope of the P_e versus δ curve at the operating point were positive. This means that the generator must be operated with a power angle within the range of 0 to 90°. Symbolically,

$$\left(\frac{d}{d\delta}P_e\right)_{\delta=\delta_0} > 0 \quad \text{condition for stable operation for small load changes}$$

2.8 Natural Frequency of Oscillation

A synchronous generator or motor in steady-state operation runs at synchronous speed (commonly 377 radians per second). However, sometimes a transient in the system, such as load changes or capacitors discharging, introduces low-frequency (subharmonic) oscillations in the electrical system that ride on top of the 60-Hz fundamental and consequently change the wave shape of the power. Fortunately these oscillations disappear quickly because of the damping effects of the electrical load, the synchronous machine itself, and the prime mover. The undamped oscillations are sinusoidal in wave shape, and Eq. (2.52) provides its angular frequency.

$$\omega_n = \sqrt{\frac{\omega_s \cdot P_{sy}}{2 \cdot H}} \tag{2.52}$$

$$\omega_n = 2 \cdot \pi \cdot f_n \qquad f_n = \frac{1}{2 \cdot \pi} \cdot \sqrt{\frac{\omega_s \cdot P_{sy}}{2 \cdot H}} \quad \text{natural frequency of oscillation}$$

$$T_n = \frac{1}{f_n} \quad \text{period of natural frequency of oscillation}$$

A machine natural frequency of oscillation is a function of its synchronizing power coefficient P_{sy} and its inertia constant H. Inertia constant H could not be negative, but P_{sy} could be, provided that cos δ_0 is negative or that the power angle is larger than 90° and smaller than 270°. If P_{sy} is negative, the natural frequency becomes complex. In general, the H values of different machines are different and consequently have different natural frequency of oscillation. It could happen that a group of coherent machines oscillates together because they have the same natural frequency of oscillation, and sometimes a group of coherent machines oscillate with respect to another group of coherent machine.

36 Chapter Two

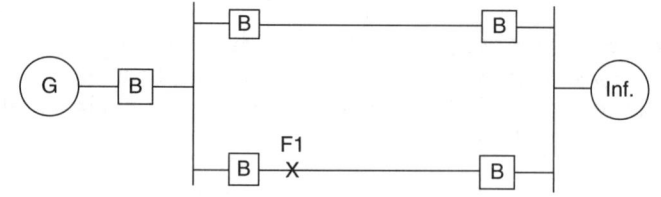

FIGURE 2-4 Network with a three-phase fault at F1.

2.9 Equal-Area Criterion of Stability

Let us consider the simple electrical network illustrated in Fig. 2-4. A three-phase symmetrical short circuit occurs at the F1 point; after a short time the fault is cleared by the two nearest breakers to point F1. *Before* the fault, the net mechanical power input to the generator's shaft P_s equals the net generator output electric power P_e, and therefore the power accelerating the rotor is zero ($P_a = P_s - P_e = 0$). In this condition the generator operates at synchronous speed ω_s, and power angle δ_0. *During* fault conditions, the generator output electric power P_e is not zero because the generator keeps delivering a reduced amount of power through the not affected transmission line, and that is good. If the fault were not three-phase symmetrical, then the generator could also keep delivering power through the not affected phases of the faulted transmission line and improve even more the chances for stability, provided that the fault-clearing breakers open only the affected poles. The power unbalance during fault conditions ($P_s > P_e$) produces an accelerating torque in the shaft (the turbine governor has not reacted yet), and the rotor increases its angular speed beyond synchronism. The fault must be cleared before the increasing power angle reaches a critical value of δ_c, which is specific for every event. The power angle plot for F1 fault is shown in Fig. 2-5, where area A_1 represents the rotor accelerating period during fault condition; this area is surrounded by P_s, δ_0, δ_c, and P_{du}. When the fault is cleared, the generator output power jumps to a point in P_{af} (after-fault power angle sine curve) and starts developing area A_2 that initiates the braking period. Here P_{af} represents the generator electric output power after the fault is cleared, and because it is higher than P_s one hopes there is enough "space" between the curve and the horizontal line representing P_s to contain an area A_2, at least equal to A_1. If this is the case, the generator will slow down to synchronous speed and fall back into stable operation. Thus the generator would be able to keep delivering power to the load still connected to the power system. Simply stated, the stability criterion is: A_1 must be equal to A_2. However, if the fault were cleared beyond δ_c, the braking period would not be long enough ($A_2 < A_1$) to reduce the angular speed to synchronous operation.

Power Systems Stability

The power angle diagram shown in Fig. 2-5 was plotted assuming the following data:

$P_s = 1.000$ net mechanical power input to generator or shaft power

$P_{be} = 2.000 \cdot \sin \delta$ generator electric output power before fault

$P_{du} = 0.5 \cdot \sin \delta$ generator electric output power during fault

$P_{af} = 1.5 \cdot \sin \delta$ generator electric output power after fault

The power angle δ_0 is provided by the first point of intersection of the horizontal line representing P_s with the curve representing P_{be}. Symbolically,

$$1.000 = 2.000 \cdot \sin \delta_0 \qquad \delta_0 := a\sin\frac{1}{2} \qquad \delta_0 = 0.524 \cdot \text{rad} \qquad \delta_0 = 30°$$

The power angle δ_{max} is provided by the second intersection of the horizontal line representing P_s with the curve representing P_{af}. Symbolically,

$$1.000 = 1.5 \cdot \sin \delta_{max} \qquad \delta_{max} := \pi - a\sin\frac{1}{1.5} \qquad \delta_{max} = 2.412 \cdot \text{rad}$$

$$\delta_{max} = 138.19°$$

$$A_1 = 1 \cdot (\delta_c - \delta_0) - \int_{\delta_0}^{\delta_c} 0.5 \cdot \sin \delta \, d\delta \qquad \cos \delta_0 = 0.866$$

$$\int_{\delta_0}^{\delta_c} 0.5 \cdot \sin \delta \, d\delta = -0.5 \cdot \cos \delta_c + 0.5 \cdot \cos \delta_0 = -0.5 \cdot \cos \delta_c + 0.5 \cdot (0.866)$$

$$A_1 = \delta_c - 0.524 + 0.5 \cdot \cos \delta_c - 0.433 = \delta_c + 0.5 \cdot \cos \delta_c - 0.957$$

$$A_2 = \int_{\delta_c}^{\delta_{max}} 1.5 \cdot \sin \delta \, d\delta - 1 \cdot (\delta_{max} - \delta_c) \qquad \cos 2.412 = -0.745446$$

$$\int_{\delta_c}^{\delta_{max}} 1.5 \cdot \sin \delta \, d\delta = -1.5 \cdot \cos \delta_{max} + 1.5 \cdot \cos \delta_c = 1.1182 + 1.5 \cdot \cos \delta_c$$

$$A_2 = 1.1182 + 1.5 \cdot \cos \delta_c - \delta_{max} + \delta_c \rightarrow A_2 = \delta_c + 1.5 \cdot \cos \delta_c - 1.2938$$

$A_1 = A_2$ stability criterion

$$\delta_c + 0.5 \cdot \cos \delta_c - 0.957 = \delta_c + 1.5 \cdot \cos \delta_c - 1.2937$$

$$0.3368 = \cos \delta_c \qquad \delta_c := a\cos 0.3368 \qquad \delta_c = 1.227 \, \text{rad} \qquad \delta_c = 70.318°$$

$$\delta := 0, \frac{\pi}{12} \cdot \pi$$

Chapter Two

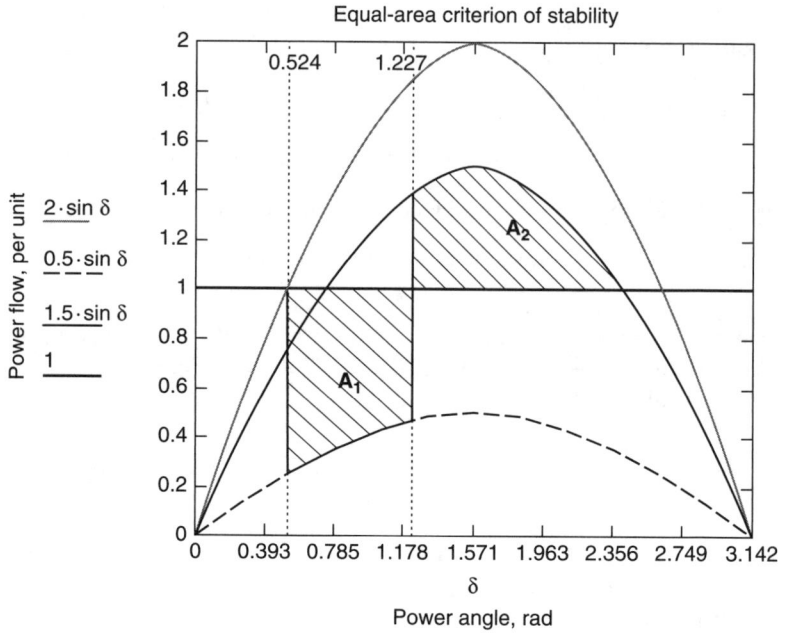

Figure 2-5 Power flow as function of δ, illustrating the equal-area criterion of stability.

$\delta_0 = 0.524$ rad — This is the initial power angle when the machine is operating synchronously and the mechanical power input is equal to the electric power output, neglecting losses.

$\delta_{max} = 2.412$ rad — This is the maximum power angle. Beyond this angle the input of mechanical power to the generator (P_s) is greater than the electric power generated by the generator after the fault has been cleared (P_{af}). Consequently, the rotor angular speed will increase beyond synchronous speed, and the generator will become unstable.

$\delta_c = 1.227$ rad — This is the critical clearing angle. When a fault occurs, the power angle starts changing and keeps increasing. To avoid exceeding, the transient stability limit of the power angle δ_{max}, the fault must be cleared before the power angle reaches the critical angle δ_c.

The graphic interpretation of a power flow diagram as a function of the power angle is enough to predict the transient stability of a generator-infinity bus system without the need to solve the swing equation. The procedure is illustrated in Fig. 2-5. This very helpful procedure is called equal-area criterion of stability. However, the solution of the swing equation is so easily accomplished using a numerical solver that the true value of the equal-area criterion is to utilize the power flow versus power angle diagram to find the values of δ_0, δ_c and δ_{max}.

2.10 Generator-Infinity Bus Network

From a stability point of view, the most damaging types of short circuit faults, in decreasing order, are

1. Three-phase symmetrical short circuit
2. Double line-to-ground short circuit
3. A line-to-line short circuit
4. A line-to-ground short circuit

The most common short circuit event is the single line to ground, and the most severe but least common is the three-phase symmetrical short circuit. Also, very rarely or never is the short circuit bolted except for intentional sabotage.

Because the rotor accelerating area A_1 decreases when the generator keeps delivering power during the fault condition, the more power it delivers, the smaller is A_1. Therefore, it is very convenient to use parallel transmission lines. Furthermore, it is also convenient to install circuit breakers for protection or reclosing duty that are capable, in case of line-to-ground faults, of opening only the affected poles.

The maximum power that could be transferred in a generator-infinity bus network without exceeding the stability limit provided by Eq. (2.15), which is adapted and repeated now:

$$P_{max} = \frac{|E_g| \cdot |E_m|}{|X|}$$

Equation (2.15) indicates the methods that could be used to improve the transient stability of the classical generator-infinity bus power system:

1. *Raise the system voltage.* Take into account that one half-wave of the voltage delivered cannot exceed the amount of volt-seconds tolerated by connected step-up transformers. Otherwise the step-up transformer cores will saturate and if they

are not tripped off the line, they will burn. In addition, the voltage delivered must not damage any connected load.

2. *Reduce the series reactance by using parallel transmission lines.* This both reduces X and increases the power angle of the critical stability limit. Also in the case of a fault in one of the transmission lines, the network will keep delivering power through the remaining line, with the net effect of increasing the system stability because A_1 will be smaller than otherwise, and shorter the accelerating period.

3. *Reduce the series reactance by using series capacitor banks.* This will increase the stability limit, but capacitor use is risky. In, the case of a short circuit in the network, the line current will contain a steady-state term and a transient term. In general, the transient term is oscillatory with a frequency that depends on the circuit parameters R, L, and C, and therefore resembles the natural frequency of oscillation of the faulted circuit. Although this frequency is usually smaller than the system synchronous frequency (60 Hz), it could also be larger. Fortunately, this transient response is damped, and its time constant is usually smaller than 100 milliseconds (ms). The difference between the system frequency and the natural frequency of the faulted circuit is called the *frequency complement*.

Actually, the transient term of the current circulating through the armature induces, at the complement frequency, currents and torques in the rotor that could cause a shaft failure or extreme vibration if the rotor natural frequency of oscillation is equal or close to the frequency complement.

4. Besides the conclusions derived from Eq. (2.15) it is essential to use high-speed circuit breakers. A fast fault interruption will decrease A_1. Furthermore, it is a plus to use breakers that open only the faulted phase. That will increase the power delivered during fault conditions and further reduce A_1.

2.11 Introduction to Stability of Multimachine Power Systems

An effort is always made to convert multimachine stability problems to the classical model of one machine feeding an infinity bus. One method of solution lumps together generators with similar characteristics and whose operating power angles are equal or almost equal into a single equivalent one. Then it assumes that the fault affecting any of the generators, lumped or not, does not affect the other generators connected to the system. In general, this approach is not possible because the generators usually are interconnected by a mesh of short transmission lines, and a fault in any line actually affects all of them. The rotors of the generators interconnected as such, swing asynchronously. "They swing to different rhythms." Actually the power system could respond to a fault in many different ways, including these:

1. The generator nearest to the fault sustains increasing synchronous oscillation before losing synchronism and is the only one to lose synchronism.
2. The generator nearest to the fault is the first to lose synchronism and is soon followed by the other generators connected to the system.
3. Some generators near the fault sustain rotor swings without losing stability. However, one or more of the generators connected to the system far away from the fault loses synchronism. It also could happen that the generator closer to the fault become unstable and loses synchronism without sustaining rotor swings. Meanwhile, some generators at longer distance from the fault location sustain synchronous oscillations, but they eventually return to stable synchronous operation.

2.12 Coherent Machines

Generators that swing together responding to system disturbances are called *coherent machines*. Coherent generators can be lumped together even if they have different speed ratings, provided that ω_s and δ are expressed in a consistent set of units. When generators are lumped together to solve transient stability problems, the single equivalent generator must have a rating equal to the sum of the ratings of the individual generators. Also, the angular momentum at synchronous speed of the equivalent machine M must be equal to the sum of the angular moment at synchronous speed of the individual machines. Once the value of M is known for the equivalent machine, the inertial constant H of the equivalent machine could be calculated using Eq. (2.22). Furthermore, in a multimachine system all the components must be expressed in a common MVA base including the inertial constant H. For instance,

Common base: 100 MVA

Generator 1 base: 80 MVA, same as generator 1 rating

Generator 1 inertial constant $H_1 = 5$ stored rotational energy, megajoules/80

$$100 \cdot H_{common} = 80 \times 5 \qquad H_{common} = \frac{400}{100} = 4 \text{ megajoules}/100$$

In general the conversion formula is

$$\text{Common base} \times H_{common} = H_g \times \text{generator base}$$
$$H_{common} = H_g \times \text{generator base/common base} \qquad (2.53)$$

In stability studies H instead of M is the preferred constant, because it has a narrow range of values for each class of machine. For instance, for 3600 rpm turbine-driven generators the H range is 4 to 7 megajoules/MVA. Generator manufacturer's sometimes provide the WR^2 which is the weight of the rotating parts in pounds times the square of the radius of gyration in feet, instead of H. To calculate H, follow the simple procedure given in Sec. 2.4, Eqs. (2.27) to (2.29).

When the swing equation is applied to several machines lumped together, the following apply:

$$H = \sum_i H_i \qquad P_s = \sum_i P_{si} \qquad P_e = \sum_i P_{ei} \qquad \delta \text{ is same for all machines}$$

2.13 Modeling of Multimachine Power Systems

Short circuits in the interconnecting network or other strong network disturbances could change the system operation from steady-state to transient. Under transient conditions, one, several, or all generators connected to the electrical system could fall into an oscillatory mode of operation. The cause of the oscillatory operation of the entire system could be a single generator that under transient conditions fell in oscillatory operation, and whose oscillations and synchronizing power transmitted through the interconnecting transmission lines affected the other generators and triggered their oscillatory operation as well. The oscillation, in general, is composed of a 60 Hz fundamental, a low-frequency harmonic close to the natural frequency of oscillation of the affected part of the system (1 to 4 Hz), and sometimes also high-frequency harmonics. Often capacitors are the source of these low- or high-frequency harmonics.

The swing curves of the machine undergoing simultaneous transient oscillations are determined by the 60 Hz and low-frequency components. Therefore, the assumption is usually made that the network parameters should be computed for 60 Hz. In addition, to decrease the amount of data and the amount of computations required in multimachine stability studies the assumptions listed below are frequently made:

1. The mechanical power input to each machine remains constant.
2. Damping power is negligible.
3. The classical model for a single machine could be used for each synchronous generator, that is, constant emf behind

constant transient reactance. Both are assumed to remain constant through the transient.

4. The mechanical rotor angle of each generator is equal to δ, which is the phase angle of the generated emf phasor with respect to the generator terminal voltage.
5. All electrical loads could be represented by shunt passive impedances connected to ground with the values they have immediately before the disturbance event that triggered the oscillations.

Classical stability models of multimachine power systems are those based in the above assumptions. Although these models are limited to first swing analysis, they are useful because they enhance our understanding of what could or did happen.

The equal-area criterion of stability is not applicable to multimachine power systems, and the three-phase symmetrical short circuit is not necessarily the worst fault that can occur. So all types of faults must be considered, especially double line-to-ground faults. Furthermore, instability studies require knowledge of the system conditions just before the fault occurs as well as the configuration of the network during and after the fault occurrence. Multimachine stability studies based on the classical model are not comprehensive enough because they do not consider the important factors listed below:

- How the system performs 2 and 3 seconds after the fault. In this case, the nonlinearity of the system components will come into play, specially, the magnetic saturation of transformers, generators, and reactors.
- The time response of the excitation system of each generator to a sudden change in its output voltage.
- How generators and turbine shafts and their couplings respond to the twisting torques produced by possible low-frequency nonharmonic oscillations when they are almost equal to the torsional natural frequency of any of the shafts.
- How fast the turbine valve controls respond to the changes produced by the fault in the generator parameters.

Figures 4-1 and 4-4 illustrate the classical model of a multimachine system.

2.14 Power Flow in a Multimachine Network

The object of power flow studies is to determine the voltage magnitude and phase angle at each bus and the flow of real and reactive power in each line. A good single-line diagram of the power system is absolutely necessary to perform the analysis. The essential information that it must contain includes all connected equipment (generators, transformers, capacitors, etc.), ratings, and impedances; and the series impedances and shunt admittances of all transmission lines. From the single-line diagram the network admittance diagram is created. Generators and loads are considered outside the network, and they are not included in the network bus admittance matrix. To determine the network's power flow pattern, it is necessary to determine the admittance of all the lines interconnecting the network's nodes (busses), and to specify at each bus either the net flow of power, real and reactive, into the network or the voltage magnitude and phase angle. At *load busses,* the most convenient selection is to specify net flow of power. At *generator busses* the most practical selection usually is to specify the magnitude and angle of the bus voltage. Rather than try to find a pure analytical solution to the problem, the standard method is to use an iterative process in which a set of estimated values are assigned (educated guess) to the unknown bus voltages. Then using the estimated set of bus voltage values and the real and reactive power specified, we calculate a new value for each bus voltage. In this way we obtain a new set of bus voltage values, and we are ready for another iteration of the procedure. The repetitive use of the procedure generates diminishing results, and it is stopped when the bus voltage changes are smaller than an established minimum value. When the voltage of load busses is changed, common practice is to maintain the reactive power flow constant. For generator busses the practice is to maintain the voltage magnitude constant.

It is impossible to specify the net flow of real power in all the busses. So one generator bus, called the *swing bus,* is left without specifying the net real power flow. The generator or generators feeding the swing bus supply the difference between the total system net real power output plus losses and the net real power flow into the system at all the other busses. The power delivered to a load is negative input power to the system. Generators and interconnection with other power systems provide positive or negative power inputs into the network.

A detailed application of the procedure is left for the example presented in Chap. 4.

2.15 Network Reduction Techniques

The simplest technique is to use wye-delta or star-mesh transformation formulas. This method works well with simple networks, but it should not be used with larger and more complicated ones. The method of node elimination by matrix algebra is presented below.

$$\mathbf{I} = \mathbf{Y} \cdot \mathbf{V} \quad \text{standard node equation in matrix notation}$$

where **I** and **V** are column matrices or vectors and **Y** is a symmetrical square matrix. The column matrices must be arranged so that the elements associated with nodes to be eliminated are in the lower rows of the matrices. The admittance matrix is partitioned so that elements associated with nodes to be eliminated are separated from the other elements by horizontal and vertical lines.

$$\begin{pmatrix} \mathbf{I}_A \\ \mathbf{I}_X \end{pmatrix} \begin{pmatrix} \mathbf{K} & \mathbf{L} \\ \mathbf{L}^T & \mathbf{M} \end{pmatrix} \cdot \begin{pmatrix} \mathbf{V}_A \\ \mathbf{V}_X \end{pmatrix} \qquad (2.54)$$

where $\mathbf{I}_X$ is the submatrix composed of the currents entering the nodes to be eliminated and $\mathbf{V}_X$ is the submatrix composed of the voltages of these nodes. Every element of $\mathbf{I}_X$ must be zero, because if not, these nodes could not be eliminated. Matrix **K** contains the self- and mutual admittances of the node to be retained. Matrix **M** is a square matrix whose order is equal to the number of nodes to be eliminated and it contains the self- and mutual admittances of the nodes to be eliminated.

Matrix **L** contains the mutual admittances common to a node to be retained and to one to be eliminated. Matrix $\mathbf{L}^T$ is the transpose of **L**. From Eq. (2.54) we obtain the following equations:

$$\mathbf{I}_A = \mathbf{K} \cdot \mathbf{V}_A + \mathbf{L} \cdot \mathbf{V}_X \qquad (2.55)$$

$$\mathbf{I}_X = \mathbf{L}^T \cdot \mathbf{V}_A + \mathbf{M} \cdot \mathbf{V}_X \qquad (2.56)$$

Since all the elements of $\mathbf{I}_X$ are zero, subtracting $\mathbf{L}^T \mathbf{V}_A$ from both sides of Eq. (2.56) and multiplying both sides by $\mathbf{M}^{-1}$ we have

$$-\mathbf{L}^T \cdot \mathbf{V}_A = \mathbf{M} \cdot \mathbf{V}_X \qquad \mathbf{M} \cdot \mathbf{M}^{-1} = \mathbf{M}^{-1} \cdot \mathbf{M} = \mathbf{U} \qquad \text{unit matrix}$$

$$-\mathbf{M}^{-1} \cdot \left(\mathbf{L}^T \cdot \mathbf{V}_A \right) = \mathbf{V}_X$$

Therefore, substituting $\mathbf{V}_X$ in Eq. (2.55), we have

$$\mathbf{I}_A = \mathbf{K} \cdot \mathbf{V}_A - \mathbf{L} \cdot \left[\mathbf{M}^{-1} \cdot \left(\mathbf{L}^T \cdot \mathbf{V}_A \right) \right] \qquad \mathbf{I}_A = \mathbf{K} \cdot \mathbf{V}_A - \mathbf{L} \cdot \mathbf{M}^{-1} \mathbf{L}^T \cdot \mathbf{V}_A$$

$$\mathbf{I}_A = (\mathbf{K} - \mathbf{L} \cdot \mathbf{M}^{-1} \cdot \mathbf{L}^T) \cdot \mathbf{V}_A$$

$\mathbf{I}_A = \mathbf{Y} \cdot \mathbf{V}_A$ This equation is a node equation where the admittance matrix is

$$\mathbf{Y} = \mathbf{K} - \mathbf{L} \cdot \mathbf{M}^{-1} \cdot \mathbf{L}^T \qquad (2.57)$$

The admittance matrix, Eq. (2.57), allows us to assemble, the circuit with the unwanted nodes eliminated.

Example 2-1

Figure 4-2 provides the per-unit admittance diagram of a power system based on 230 (kV) and 300 MVA. Find the equivalent network with nodes (busses) 2, 3, and 6 eliminated. These nodes could be eliminated because the loads connected to them are considered to be constant. Actually, we are assuming that these constant loads are not affected and do not participate in the dynamics of any disturbance occurring somewhere else in the network.

Nodes to be eliminated: 2, 3, 6
Nodes to be retained: 1, 4, 5

Equation (2.57) allows us to reassemble the circuit with nodes 2, 3, and 6 eliminated:

$$Y = \begin{pmatrix} K & L \\ L^T & M \end{pmatrix}$$ To eliminate nodes, admittance matrix is partitioned into four submatrices.

where **K** is a 3 × 3 square matrix composed of the self- and mutual admittances of the nodes to be *retained*. Each element of the principal diagonal of **K** is equal to the sum of all the admittances terminating on the node identified in the matrix by repeated subscripts.

$$K = \begin{pmatrix} Y_{1,1} & Y_{1,4} & Y_{1,5} \\ Y_{4,1} & Y_{4,4} & Y_{4,5} \\ Y_{5,1} & Y_{5,4} & Y_{5,5} \end{pmatrix}$$

From Fig. 4-2 we obtain all the admittances:

$Y_{1,1} := 1.248 - 7.8j$ $Y_{1,4} := 0$ $Y_{1,5} := 0$

$Y_{4,1} := 0$ $Y_{4,4} := 0.974 - 6.272j$ $Y_{4,5} := 0$

$Y_{5,1} := 0$ $Y_{5,4} := 0$ $Y_{5,5} := 0.974 - 6.272j$

$$K := \begin{pmatrix} 1.248 - 7.8j & 0 & 0 \\ 0 & 0.974 - 6.272j & 0 \\ 0 & 0 & 0.974 - 6.272j \end{pmatrix}$$ nodes to be retained: 1, 4, 5

Matrix **M** is a square matrix whose order is equal to the number of nodes to be eliminated (3). Matrix **M** is composed of the self- and mutual admittances of the nodes to be eliminated. Any element of the principal diagonal is equal to the sum of all admittances connected to the node. Each element *not* in the diagonal is equal to the negative sum of all admittances connected directly between the pair of nodes identified by the double subscript. Nodes to be eliminated: 2, 3, 6.

$$M = \begin{pmatrix} Y_{2,2} & Y_{2,3} & Y_{2,6} \\ Y_{3,2} & Y_{3,3} & Y_{3,6} \\ Y_{6,2} & Y_{6,3} & Y_{6,6} \end{pmatrix}$$

$Y_{2,2} := 1.04 - 5.897j$ $\qquad Y_{2,3} := 0 \qquad\qquad Y_{2,6} := 0$

$Y_{3,2} := 0 \qquad\qquad\quad Y_{3,3} := 1.04 - 5.897j \qquad Y_{3,6} := 0$

$Y_{6,2} := 0 \qquad\qquad\quad Y_{6,3} := 0 \qquad\qquad Y_{6,6} := 1.116 - 8.55j$

$$M := \begin{pmatrix} 1.04 - 5.897j & 0 & 0 \\ 0 & 1.04 - 5.897j & 0 \\ 0 & 0 & 1.116 - 8.55j \end{pmatrix}$$ nodes to be eliminated: 2, 3, 6

$$M^{-1} = \begin{pmatrix} 0.029 + 0.164j & 0 & 0 \\ 0 & 0.029 + 0.164j & 0 \\ 0 & 0 & 0.015 + 0.115j \end{pmatrix}$$

Matrix **L** is composed of only those *mutual* admittances common to a node to be retained and to one to be eliminated and each element is equal to the negative of the admittance between nodes.

Matrix **L** is a square matrix of the same order as the number of nodes to be retained.

Nodes to be retained: 1, 4, 5
Nodes to be eliminated: 2, 3, 6

$$L = \begin{pmatrix} Y_{1,2} & Y_{1,3} & Y_{1,6} \\ Y_{4,2} & Y_{4,3} & Y_{4,6} \\ Y_{5,2} & Y_{5,3} & Y_{5,6} \end{pmatrix}$$

$$L := \begin{pmatrix} -0.624 + 3.9j & -0.624 + 3.9j & 0 \\ -0.416 + 1.997j & 0 & -0.558 + 4.275j \\ 0 & -0.416 + 1.997j & -0.558 + 4.275j \end{pmatrix}$$

$$L^T = \begin{pmatrix} -0.624 + 3.9j & -0.416 + 1.997j & 0 \\ -0.624 + 3.9j & 0 & -0.416 + 1.997j \\ 0 & -0.558 + 4.275j & -0.558 + 4.275j \end{pmatrix} \qquad Y := K - L \cdot M^{-1} \cdot L^T$$

$$Y = \begin{pmatrix} 0.507 - 2.643j & -0.253 + 1.321j & -0.253 + 1.321j \\ -0.253 + 1.321j & 0.532 - 3.459j & -0.279 + 2.138j \\ -0.253 + 1.321j & -0.279 + 2.138j & 0.532 - 3.459j \end{pmatrix}$$

$$Y = \begin{pmatrix} Y_{1,1} & Y_{1,4} & Y_{1,5} \\ Y_{4,1} & Y_{4,4} & Y_{4,5} \\ Y_{5,1} & Y_{5,4} & Y_{5,5} \end{pmatrix}$$

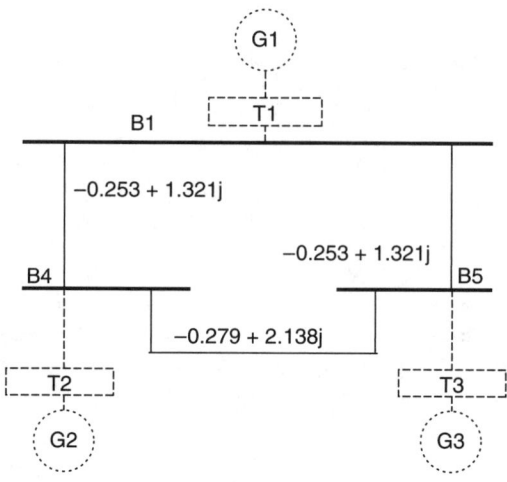

FIGURE 2-6 Per-unit admittance diagram of the power system with three nodes eliminated. Base of 230 kV and 300 MVA.

Nodes to be retained: 1, 4, 5

$Y_{1,1} := 0.507 - 2.643j$ $Y_{1,4} := -0.253 + 1.321j$ $Y_{1,5} := -0.253 + 1.321j$

$Y_{4,1} := -0.253 + 1.321j$ $Y_{4,4} := 0.532 - 3.459j$ $Y_{4,5} := -0.279 + 2.138j$

$Y_{5,1} := -0.253 + 1.321j$ $Y_{5,4} := -0.279 + 2.138j$ $Y_{5,5} := 0.532 - 3.459j$

The resulting network after eliminating nodes 2, 3, and 6, which are load busses, is illustrated in Fig. 2-6.

In the original bus admittance diagram of the network (see Fig. 4-2), generators and loads were considered outside the network, and therefore they were not included in the network bus admittance matrix Y_{bus}. See Y_{bus} in Chap. 4. There are not admittances connected between any of the busses and ground (the reference node). The impedances connected directly between nodes in the resulting circuit are the negative of the inverse of the mutual admittances. Therefore, the impedances between nodes 1, 4, 5 are

$$Z_{1,4} := -(Y_{1,4})^{-1} = 0.14 + 0.73j \quad \text{per unit}$$
$$Z_{4,5} := -(Y_{4,5})^{-1} = 0.06 + 0.46j \quad \text{per unit}$$
$$Z_{5,1} := -(Y_{5,1})^{-1} = 0.14 + 0.73j \quad \text{per unit}$$

CHAPTER 3

Transient Stability Problem in a Simple Electrical Network

In this chapter we solve a stability problem in a simple electrical network composed of a generator feeding an ideal infinity bus by way of two parallel transmission lines. First we use the equal-area criterion of stability to determine the critical power angle, and then we solve the swing equation using a numerical solver to determine the time it takes for the increasing power angle to reach the critical power angle.

3.1 Stability Problem

Figure 3-1 shows an electrical network consisting of a generator feeding an infinity bus through parallel transmission lines. The generator is delivering to the infinity bus 1.00 per-unit (pu) power when a three-phase symmetrical fault occurs at point F. The fault is cleared by the two closest breakers tripping simultaneously.

Data:

E_g = 1.25 pu constant

E_b = 1.00 pu constant

P_s = 1.00 pu constant, shaft power or mechanical input from prime mover

P_e = 1.00 pu electric power developed by generator

H = 4 MJ/MVA generator inertial constant

X'_d = 0.12j pu generator transient reactance included in value given of 0.2j

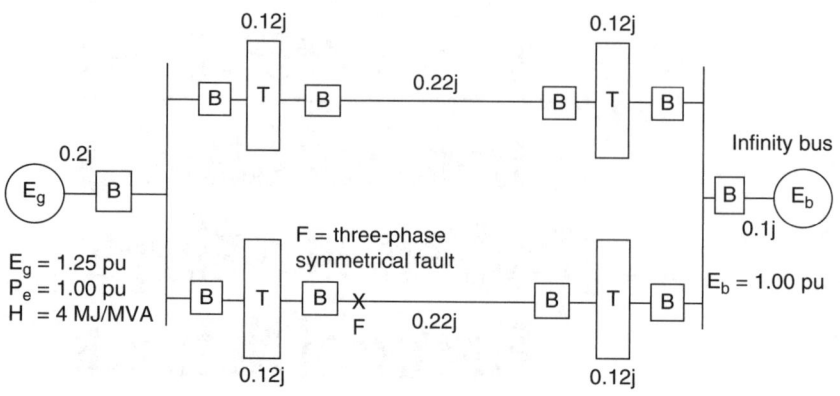

FIGURE 3-1 Single-line diagram of a simple electrical network.

3.2 Network Reduction

The electric power transmitted before, during, and after the fault is a function of the power angle δ. Figure 3-2 shows the impedance diagram of the electrical network. Next, we deduce and plot the equations of the electric power transmitted.

Before the fault, the reactance between the generator and the infinity bus is

$$0.2 + \frac{0.12 + 0.22 + 0.12}{2} + 0.1 = 0.53 \qquad x_b := 0.53j$$

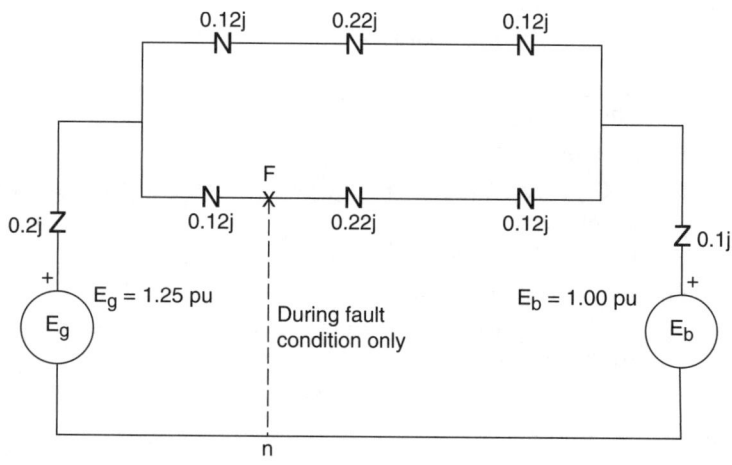

FIGURE 3-2 Impedance diagram of power system.

Transient Stability Problem in a Simple Electrical Network

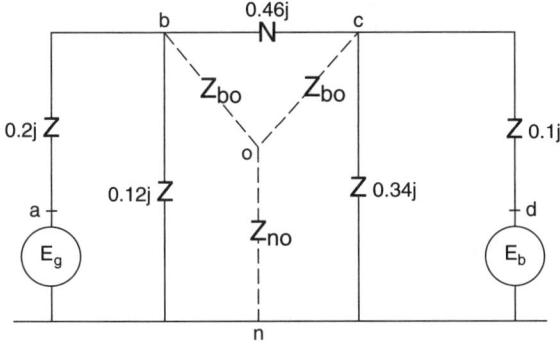

FIGURE 3-3 Network during three-phase symmetrical fault.

After the fault is cleared by the circuit breakers at both ends (see Fig. 3-1), the reactance between the generator and the infinity bus is

$$0.2 + 0.12 + 0.22 + 0.12 + 0.1 = 0.76 \quad X_a := 0.76j$$

During the three-phase *symmetrical* fault, the potential of the point at which the fault occurs is the same as the potential of the generator neutral. Figure 3-3 shows the network during fault conditions. Transforming the central delta into a wye, we have

$$\frac{0.12 \times 0.46}{0.12 + 0.46 + 0.34} = 0.06 \quad X_{bo} := 0.06 \cdot j$$

$$\frac{0.46 \times 0.34}{0.12 + 0.46 + 0.34} = 0.17 \quad X_{co} := 0.17 \cdot j$$

$$\frac{0.12 \times 0.34}{0.12 + 0.46 + 0.34} = 0.044 \quad X_{no} := 0.0443 \cdot j$$

After the delta-wye transformation the network looks as shown in Fig. 3-4.

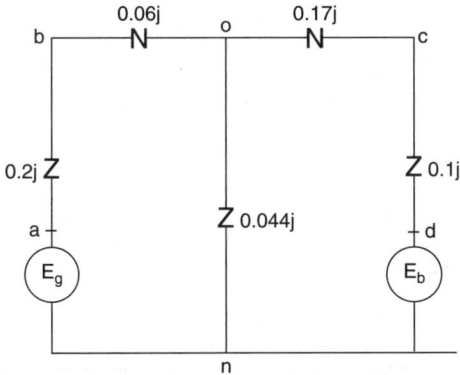

FIGURE 3-4 Network during fault conditions after delta-wye transformation.

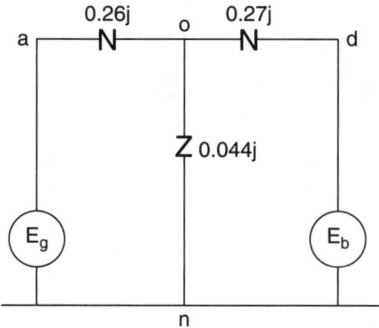

FIGURE 3-5 Network after combining series reactances.

Figure 3-5 is obtained by combining series reactance as follows:

$$0.2 \cdot j + 0.06 \cdot j = 0.26j \qquad 0.1 \cdot j + 0.17 \cdot j = 0.27j$$

$$X_{ao} := 0.26 \cdot j \qquad X_{no} := 0.044 \cdot j \qquad X_{do} := 0.27j$$

The final equivalent delta network of the original system during a three-phase symmetrical fault condition is shown in Fig. 3-6. The delta-wye transformation is as follows:

$$0.26j \cdot 0.27j + 0.27j \cdot 0.044j + 0.044j \cdot 0.26j = -0.0935$$

$$X_{ad} := \frac{-0.0935}{0.044j} = 2.125j \qquad X_{dn} := \frac{-0.0935}{0.26j} = 0.3596j$$

$$X_{na} := \frac{-0.0935}{0.27j} = 0.3463j$$

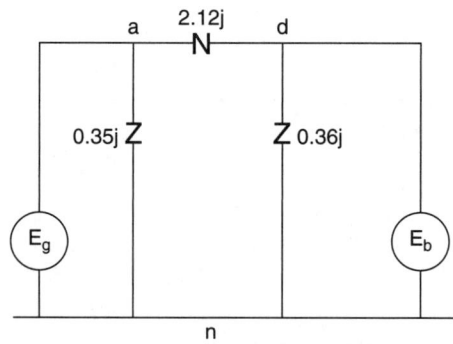

FIGURE 3-6 Equivalent delta network during fault condition.

3.3 Electric Power Transmitted

$E_g := 1.25$ $\quad\quad E_b := 1.00$

$P_e = \dfrac{|E_g| \cdot |E_b|}{|x|} \cdot \sin(\delta)$

Generator electrical power output or electrical power flow between the generator and the infinity bus; δ is the power or torque angle in electrical radians.

$P_{emax} = \dfrac{|E_g| \cdot |E_b|}{|x|}$

Maximum possible flow of electrical power. Resistances were neglected, hence, there are no copper losses, and the infinity bus receives all the electrical power generated.

$P = P_{max} \sin \delta$

In steady-state condition, power angle is not changing and electric power transferred cannot be larger than mechanical power delivered by drive shaft.

Before the Fault

Before the fault the generator maximum electric output power is

$$P_{max} := \dfrac{|E_g| \cdot |E_b|}{|X_b|} = 2.3585$$

The per-unit power transferred before the fault is

$$P_{be} = 2.385 \cdot \sin \delta \quad\quad (3.1)$$

The electric power transferred cannot be larger than the mechanical power delivered by the driveshaft. Actually, in synchronous operation the generator output electric power must be equal to the mechanical input power delivered by the shaft minus losses.

$P_s := 1.00$ constant mechanical input power from prime mover in per unit

$1.00 = 2.385 \cdot \sin \delta_0$ first intersection

$a \sin\left(\dfrac{1}{2.385}\right) = 0.4327 \text{ rad} \quad\quad a \sin\left(\dfrac{1}{2.385}\right) = 24.7896°$

$\delta_0 := 0.4327 \text{ rad}$

During Fault Conditions

When a fault occurs in the electrical power system, the voltage at all busses decreases, including the generator bus, with the exception of the voltage at the infinity bus. However, for convenience it is assumed that the voltage E_g remains constant. So the maximum power that could be transferred during fault conditions is

$$\frac{|E_g| \cdot |E_b|}{|X_{ad}|} = 0.59$$

The impedances X_{an} and X_{dn} connected across the generator and infinity bus, as shown in Fig. 3-6, are pure inductances and cannot absorb real power. So the only reactance to be considered in the computation of the power transferred during fault conditions is $X_{ad} = 2.12j$.

The per-unit electric power transferred during fault conditions is

$$P_{du} = 0.59 \cdot \sin \delta \tag{3.2}$$

After the Fault

The maximum power that could be transferred after the fault is cleared by the two closest breakers, one in each side, is

$$\frac{|E_g| \cdot |E_b|}{|X_a|} = 1.6447$$

The per-unit electric power transferred after fault is cleared is

$$P_{af} = 1.6447 \sin \delta \tag{3.3}$$

$P_s = 1.6447 \sin \delta$ Equating shaft power to the transferred electric power.

$$1 = 1.6447 \sin \delta \quad a \sin\left(\frac{1}{1.6447}\right) = 37.446° \quad \delta = 0.6536 \text{ rad} = 37.45°$$

The power angle at the second intersection of the P_{af} curve with the horizontal line representing the shaft power is called δ_{max} (see Fig. 3-7). Beyond this point the electric power transferred becomes smaller than the shaft power.

$$\delta_{max} := \pi - 0.6536 = 2.488 \text{ rad} \quad 180° - 37.446° = 142.554°$$

$$\delta_{max} = 2.488 \text{ rad} = 142.554°$$

Transient Stability Problem in a Simple Electrical Network

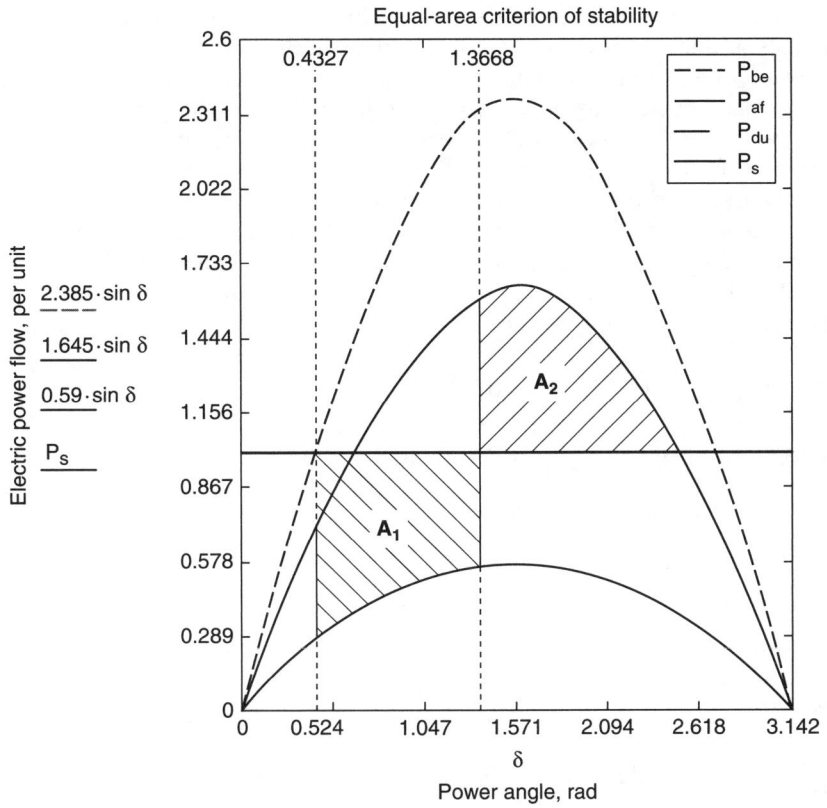

Figure 3-7 Criterion of transient stability for one-generator, one-infinity-bus network.

3.4 Power Transmitted Before, During, and After Fault Conditions

The per-unit power transmitted before, during, and after the fault are plotted in Fig. 3-7 as a function of the power angle δ. The mechanical input power from the generator's prime mover P_s is assumed constant at 1.00 per unit and all losses (frictional, wind, copper, and iron) are neglected.

$$P_s = 1.00 \quad P_{be} = 2.385 \cdot \sin \delta \quad P_{du} = 0.59 \cdot \sin \delta$$

$$P_{af} = 1.645 \cdot \sin \delta$$

$$\delta = 0.4327 \text{ rad} \quad \delta_{max} = 2.488 \text{ rad} \quad \delta_c = 1.3668 \text{ rad}$$

δ_0 is the power angle when the machine is operating synchronously and just before the disturbance. The mechanical input power is equal to the electric output power, neglecting losses.

δ max beyond this power angle the input mechanical power to the generator is greater than the electric power generated by the generator; consequently, its angular speed will increase beyond synchronous speed, and the generator will become unstable.

δ_c is the critical clearing angle. When a fault occurs, the power angle starts to increase, and to avoid exceeding the transient stability limit of the power angle δ_{max} the fault must be cleared before the power angle reaches the critical angle δ_c.

$$A_1 = 1(\delta_c - \delta_0) - \int_{\delta_0}^{\delta_c} 0.59 \cdot \sin \delta \, d\delta$$

$$\int_{\delta_0}^{\delta_c} 0.59 \cdot \sin \delta \, d\delta \to \int_{0.4327 \text{ rad}}^{\delta_c} 0.59 \cdot \sin \delta \, d\delta \to -0.59 \cdot \cos \delta_c + 0.59 \cdot \cos(0.4327 \text{ rad})$$

$$A_1 = (\delta_c - 0.4327) - [-0.59 \cdot \cos \delta_c + 0.59 \cdot \cos(0.4327 \text{ rad})]$$
$$A_1 = \delta_c - 0.4327 + 0.59 \cdot \cos \delta_c - 0.59 \times 0.9078$$
$$A_1 = \delta_c + 0.59 \cdot \cos \delta_c - 0.9683$$

$$A_2 = \int_{\delta_c}^{\delta_{max}} 1.645 \cdot \sin \delta \, d\delta - 1(\delta_{max} - \delta_c)$$

$$\int_{\delta_c}^{\delta_{max}} 1.645 \cdot \sin \delta \, d\delta \to \int_{\delta_c}^{2.488} 1.645 \cdot \sin \delta \, d\delta \to 1.645 \cdot \cos \delta_c + 1.306$$

$$A_2 = 1.645 \cdot \cos \delta_c + 1.306 - (2.488 - \delta_c)$$
$$A_2 = \delta_c + 1.645 \cdot \cos \delta_c - 1.182$$

Criterion of Stability: $A_1 = A_2$

$$\delta_c + 0.59 \cdot \cos \delta_c - 0.9683 = \delta_c + 1.645 \cdot \cos \delta_c - 1.182$$

$$0.2137 = 1.055 \cdot \cos \delta_c \qquad \delta_c = a \cos \frac{0.2137}{1.055} \to \delta_c = 1.367$$

$$\delta_c := 1.367 \text{ rad} = 78.3233°$$

This critical clearing angle is shown in Fig. 3-7.

3.5 Swing Equation

To select the protection equipment, we need to express δ as a function of time.

$$\frac{H}{\pi \cdot f} \cdot \frac{d^2}{dt^2} \delta(t) = P_{s,pu} - P_{e,pu} \qquad \text{swing equation} \qquad \text{See Eq. (2.37).}$$

Transient Stability Problem in a Simple Electrical Network

where $P_{s,pu}$ = per-unit shaft power or input from the prime mover; losses neglected
$P_{e,pu}$ = per-unit generator developed electric power
H = inertial constant in megawatt-seconds per MVA of generator rating

$$M = I \cdot \omega \quad \frac{\text{megajoule-second}}{\text{radian}} \quad \text{See Table 2-1.}$$

M is the angular momentum expressed in megajoule-seconds per radian (mJ·s/rad).

$$\omega = \frac{\text{radian}}{\text{second}} \qquad I = \frac{M}{\omega} = \frac{\text{megajoule-second}^2}{\text{radian}^2}$$

The stored rotational energy at synchronous speed is

$$\frac{I \cdot \omega^2}{2} = \frac{M \cdot \omega}{2} \quad \text{megajoule}$$

The inertial constant H is the stored rotational energy of the machine at synchronous speed in per unit of the generator rating in MVA. Symbolically,

H = stored rotational energy in megajoules/generator MVA rating

$$H = \frac{M \cdot \omega}{2 \cdot G} \quad G = \text{generator rating, MVA} \quad G \cdot H = \frac{M \cdot \omega}{2} \quad \text{megajoule}$$

GH = stored rotational energy, in megajoule

$$\omega = 2\pi \cdot f \qquad \frac{\omega}{2} = \pi \cdot f \qquad G \cdot H = M \cdot \pi \cdot f \quad \text{megajoule}$$

$$M = \frac{G \cdot H}{\pi \cdot f}$$

H = stored rotational energy in per unit of generator MVA rating

For $G := 1$ per-unit value $\quad M = \dfrac{H}{\pi \cdot f}$

Equation (2.37) provides the swing equation:

$$\frac{H}{\pi \cdot f} \cdot \frac{d^2}{dt^2} \delta(t) = P_{s,pu} - P_{e,pu}$$

For the specific case illustrated in Fig. 3-1, H = 4 and f = 60. From Eq. (3.2) we obtain the per-unit electric power flowing during the fault. Symbolically,

$$P_d = 0.59 \sin \delta$$

$$M = \frac{4}{60 \cdot \pi} = 0.0212 \qquad 0.0212 \cdot \frac{d^2}{dt^2}\delta(t) = 1 - 0.59 \cdot \sin \delta$$

$$\frac{d^2}{dt^2}\delta(t) = \frac{1}{0.0212} - \frac{0.59 \cdot \sin \delta(t)}{0.0212} \quad \text{second-order differential equation}$$

(3.4)

The equal-area criterion of stability provides the critical power angle expressed in radians or degrees, but we need to know how long it takes to reach δ_c in milliseconds, which is all the time the protecting system has to clear the fault. To determine this, we must solve Eq. (3.4), the swing equation. For convenience we will use a numerical routine.

3.6 Numerical Solver

To solve a differential equation using a numerical routine, the differential equation must be an argument of the solver routine. To do this, we must put the differential equation in standard form. The accepted format is to write the equation with the derivatives alone on the left-hand side of the equals sign and to eliminate all higher-order derivatives and replace them with only first derivatives. A second-order derivative is a first derivative of a first derivative, so the swing equation can be expressed as

$$\frac{d^2}{dt^2}\delta(t) = \frac{d}{dt} \cdot \frac{d}{dt}\delta(t) \qquad \frac{d}{dt} \cdot \frac{d}{dt}\delta(t) = \frac{1}{0.0212} - \frac{0.59 \cdot \sin \delta_0(t)}{0.0212}$$

Step 1: Standardization
Define two new functions $\delta_0(t)$ and $\delta_1(t)$ where

$$\delta_0(t) = \delta(t) \qquad \text{and} \qquad \delta_1(t) = \frac{d}{dt}\delta_0(t)$$

Now the original differential equation can be written as:

$$\frac{d}{dt}\delta_1(t) = \frac{1}{0.0212} - \frac{0.59 \cdot \sin \delta_0(t)}{0.0212}$$

Transient Stability Problem in a Simple Electrical Network

This new differential equation has two functions, $\delta_0(t)$ and $\delta_1(t)$, instead of one, $\delta(t)$. These two new functions are related by the following equation:

$$\delta_1(t) = \frac{d}{dt}\delta_0(t)$$

The original differential equation has been converted to a system of two differential equations:

$\frac{d}{dt}\delta_0(t) = \delta_1(t)$ Equations have been written with derivatives alone on left-hand side of equals sign.

$$\frac{d}{dt}\delta_1(t) = \frac{1}{0.0212} - \frac{0.59 \cdot \sin \delta_0(t)}{0.0212}$$

Step 2: Single Vector Function

Define a single vector function D containing the two new functions of t for the numerical solver.

$$D(t, \delta) = \begin{pmatrix} \frac{d}{dt}\delta_0 \\ \frac{d}{dt}\delta_1 \end{pmatrix}$$

$$D(t, \delta) := \begin{bmatrix} \delta_1 \\ \frac{1}{0.0212} - \frac{0.59 \cdot \sin \delta_0(t)}{0.0212} \end{bmatrix}$$

The function D is an indispensable tool for numerically solving differential equations. When you are writing this function, use the MathCad array subscript to indicate the derivative order.

The function Rkadapt(ic, t1, t2, npoints, D) is used to solve the differential equations. This function returns a matrix of solution values for the differential equation specified by the derivatives in D and having initial conditions ic on the interval (t_1, t_2) using an adaptive Runge-Kutta method. The npoints parameter controls the number of rows in the matrix output. The initial value of δ is evaluated at t_1.

$$\delta_0 := 0.4327 \qquad t := 0 \qquad \frac{1}{0.0212} - \frac{0.59 \cdot \sin 0.4327}{0.0212} = 35.5$$

Chapter Three

Initial conditions:

$$ic := \begin{pmatrix} 0 \\ 35.5 \end{pmatrix} \qquad t1 := 0 \qquad t2 := 1 \qquad \text{npoints} := 1000$$

$$Q := \text{Rkadapt}(ic, t1, t2, \text{npoints}, D)$$

$$\delta_0 = 0.4327 \text{ rad} \qquad \delta_c = 1.367 \text{ rad} \qquad \delta_{max} = 2.488$$

$$\delta_0(t) = \delta(t) \qquad \delta_1(t) = \frac{d}{dt}\delta_0(t) \qquad \frac{d}{dt}\delta_0(t) = \delta_1(t)$$

$$\frac{d}{dt}\delta_1(t) = \frac{1}{0.0212} - \frac{0.59 \cdot \sin \delta_0(t)}{0.0212}$$

$$Q^{(2)} = \frac{1}{0.0212} - \frac{0.59}{0.0212} \cdot \sin \delta_0(t)$$

$$MQ^{(2)} = P_s - P_d \qquad \text{swing equation}$$

$$M := 0.0212 \qquad MQ^{(2)} \qquad \text{accelerating power}$$

Figure 3-8 provides the value of the power angle, 0.53 rad (30.367°) at the time of fault clearing, or 50 milliseconds (ms). If the generator remains in service after the fault is cleared, then it is necessary to determine if the system would remain stable operating in this new configuration. Table 3-1 and Fig. 3-8 provide the initial conditions for this new configuration.

$$\delta := 0.53 \text{ rad or } 30.367° \qquad Q^{(1)} = 1.8334 \text{ rad/s} \qquad Q^{(2)} = 37.8163 \text{ rad/sec}^2$$

From Eq. (3.3) we get the power transferred after the fault is cleared:
$P_{af} = 1.645 \cdot \sin\delta$

The mechanical power input from the prime mover P_s is assumed constant at 1.00 per unit, and all losses are neglected. In these conditions the function D vector becomes

$$D(t, \delta) := \begin{bmatrix} \delta_1 \\ \frac{1}{0.0212} - \frac{1.645 \cdot \sin \delta_0(t)}{0.0212} \end{bmatrix}$$

The initial conditions after the fault is cleared are determined below.

$$\delta_0 := 0.53 \qquad \frac{1}{0.212} - \frac{1.645 \cdot \sin 0.53}{0.0212} = 7.9433$$

$$ic := \begin{pmatrix} 1.8334 \\ 7.9433 \end{pmatrix} \qquad t1 := 0 \qquad t2 := 1 \qquad \text{npoints} := 1000$$

$$Q_a := \text{Rkadapt}(ic, t1, t2, \text{npoints}, D)$$

$$Q = \begin{array}{|c|c|c|c|}
\hline
 & t & \delta_1 & \frac{d}{dt}\delta_1 \\
\hline
 & 0 & 1 & 2 \\
\hline
49 & 0.049 & 1.7956 & 37.7717 \\
50 & 0.05 & 1.8334 & 37.8163 \\
51 & 0.051 & 1.8713 & 37.8609 \\
52 & 0.052 & 1.9092 & 37.9054 \\
53 & 0.053 & 1.9471 & 37.9497 \\
54 & 0.054 & 1.9851 & 37.994 \\
55 & 0.055 & 2.0231 & 38.0381 \\
56 & 0.056 & 2.0611 & 38.0821 \\
57 & 0.057 & 2.0992 & 38.126 \\
58 & 0.058 & 2.1374 & 38.1698 \\
59 & 0.059 & 2.1756 & 38.2135 \\
60 & 0.06 & 2.2138 & 38.257 \\
61 & 0.061 & 2.2521 & 38.3005 \\
62 & 0.062 & 2.2904 & 38.3438 \\
63 & 0.063 & 2.3288 & 38.3869 \\
64 & 0.064 & 2.3672 & \ldots \\
\hline
\end{array}$$

*Only 16 points are shown in Table 3-1, but in MathCad you could scroll down to any of the 1000 points.

TABLE 3-1 Solution Matrix*

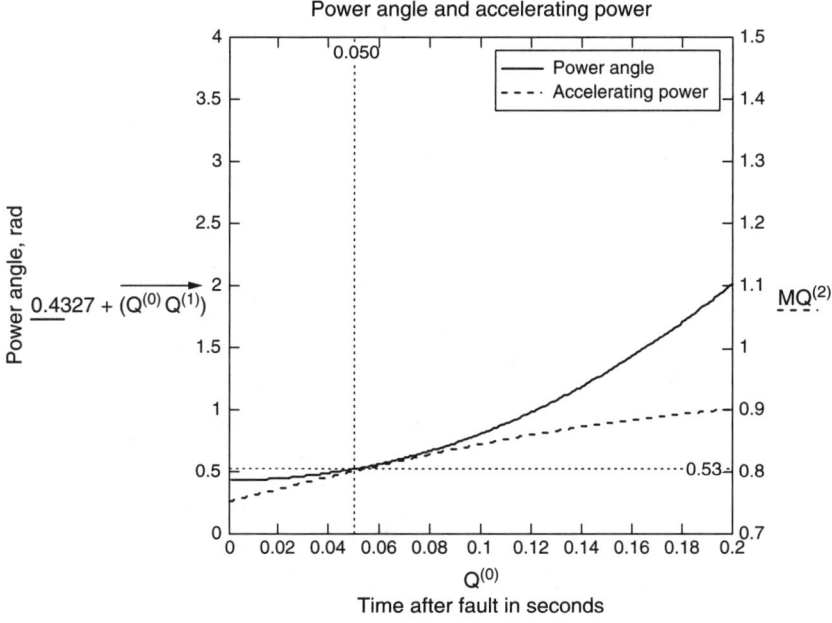

FIGURE 3-8 Power angle and accelerating power as a function of time.

		0	1	2
	0	0	1.8334	7.9433
	1	0.001	1.8414	7.9904
	2	0.002	1.8494	8.0374
	3	0.003	1.8574	8.0842
	4	0.004	1.8655	8.1308
	5	0.005	1.8737	8.1773
	6	0.006	1.8819	8.2237
$Q_a =$	7	0.007	1.8902	8.2699
	8	0.008	1.8984	8.316
	9	0.009	1.9068	8.3619
	10	0.01	1.9152	8.4077
	11	0.011	1.9236	8.4533
	12	0.012	1.9321	8.4987
	13	0.013	1.9406	8.544
	14	0.014	1.9492	8.5892
	15	0.015	1.9578	...

TABLE 3-2 Different Initial Conditions

With these new initial conditions the Rkadapt function generates Table 3-2, which provides the data to plot Fig. (3-9).

The dotted curve in Fig. 3-9 represents the swing equation, which is the accelerating power in per unit. The graphic shows that the system becomes unstable after the fault is cleared. In fact, after 400 milliseconds, the accelerating power and the power angle increase very fast. Obviously it is necessary to decrease the accelerating power after clearing the fault. The brief discussion below provides some guidelines.

Megajoule (MJ) = megawatt · second (MW · s)

$$M = \frac{\text{megajoule-second}}{\text{rad}} = \frac{\text{megawatt-second}^2}{\text{rad}}$$

$$P_a = M \cdot \alpha \quad P_a = \frac{\text{megawatt-second}^2}{\text{rad}} \cdot \frac{\text{rad}}{\text{sec}^2} = \text{megawatt}$$

Equation (2.37), the swing equation, is repeated here.

$$\frac{H}{\pi \cdot f} \cdot \frac{d^2}{dt^2} \delta(t) = P_{s,pu} - P_{e,pu} \quad \text{swing equation}$$

$$P_{s,pu} - P_{e,pu} = P_a \quad \text{accelerating power}$$

Transient Stability Problem in a Simple Electrical Network

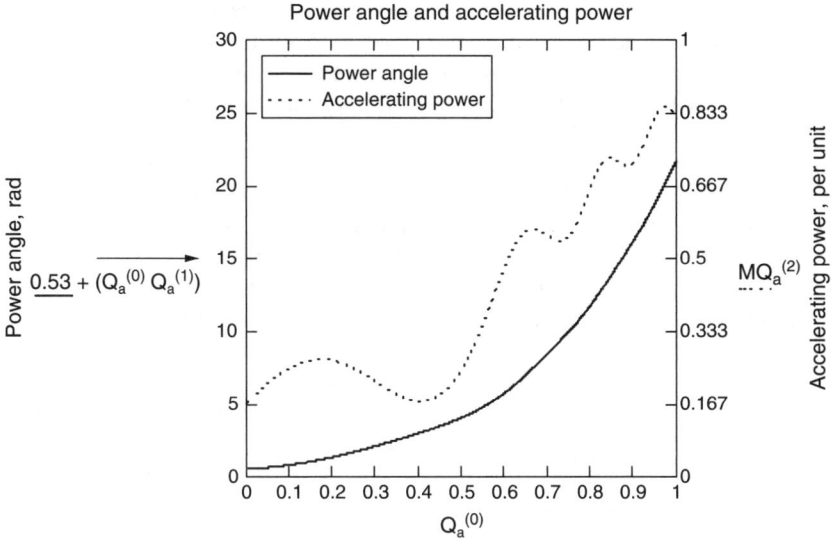

FIGURE 3-9 Power angle and accelerating power after the fault is cleared.

The shaft power is considered constant, and it should not be used to control the accelerating power, because the driver rpm should be maintained within a narrow range. The best way to decrease the accelerating power is to increase the electric power delivered by the generator during and after fault conditions. The electric power delivered by the generator after the fault is

$$P = P_{max} \cdot \sin \delta$$

where

$$P_{max} = \frac{|E_g| \cdot |E_b|}{|X_{ad}|} = 1.645 \quad \text{after clearing fault}$$

The choices to increase the electric power are

1. Increase the magnitude of the generated voltage E_g.
2. Decrease the magnitude of the reactance X_{ad} between the generator and the infinity bus. This action also has an impact on the magnitude of E_b.

Increasing the generator field excitation increases E_g. The best way to decrease X_{ad} (if you are in the design phase) is to add another transmission line. This increases both the power flow after the fault has been cleared and the reliability of the power system. A different approach for reducing the line reactance is to minimize the transformer reactance; this can only be done before ordering the transformer. Unfortunately, this solution will also increase the available short circuit currents, so it is necessary to verify that the breakers can interrupt the new and higher short circuit currents. Another approach is to add in series capacitance, but this solution will introduce a new set of problems that should be considered very carefully.

In conclusion, the system cannot operate with one line out of service, because it could become unstable following a small disturbance. So either increase the power delivered by the generator after clearing the fault or repair the faulty line before energizing the system again.

CHAPTER 4
Transient Stability Problem in a Multimachine Network

In general, the solution of a multimachine transient stability problem requires taking the following steps:

1. Collect all the necessary data.
2. Obtain a single-line diagram of the entire network showing the transmission line impedances and all the apparatus connected to the network.
3. Obtain the most recent load flow study of the network.
4. Compute the power angles of all generators connected to the network during normal operation.
5. Determine which generators are coherent, and select the ones that should be lumped together.
6. Determine the network configurations during and after fault conditions, for different locations and type of faults.
7. Create the reactance diagram of each network configuration. Select the configurations to be analyzed in the stability study.
8. Reduce the network, using node elimination techniques (matrix algebra) or the classical star-mesh transformation formulas.
9. Use a numerical solver to determine the network response to the transient disturbances.

A power system is *transient stable* if it remains in synchronism when submitted to large disturbances. The resultant large power and the generator's voltage angle oscillations prevent the linearization of the swing equations of the affected generators. In this case the study of the power system is carried out using time-domain methods of solution, such as numerical integration techniques. In simple power systems, such as one generator connected to an infinite bus or two directly connected generators, the equal-area criterion is a graphical method that provides a quick solution of the problem.

Figure 4-1 illustrates the single-phase diagram of a simple, balanced electrical network consisting of three generators and three load busses. Figure 4-2 shows the one-line admittance diagram of the same system, which contains a total of nine busses and in which all admittances are in per unit of 230 kilovolts (kV) and 300 megavolt-amperes (MVA). Table 4-1 shows type and factory rating of the three generators. Table 4-6 shows the estimated per-unit real and reactive power input to the network, as well as the bus voltages before starting the first iteration of the power system load flow study during normal operating conditions and neglecting losses. Input power, real or reactive, at any bus is considered positive. The power, real or reactive, delivered to any load is considered negative power. The reader must keep in mind that in normal operating conditions bus voltages do not differ too much in magnitude and phase angle.

Solution assumptions:

- Generators are represented by a constant voltage source behind the direct-axis transient reactance.
- The mechanical rotor angle of each generator is equal to its electrical power angle δ.

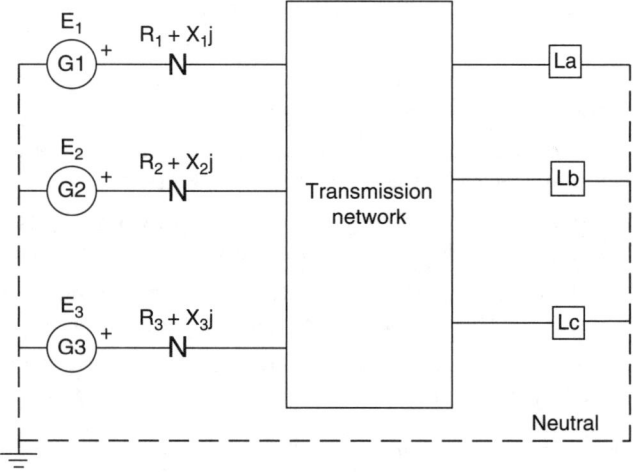

Figure 4-1 Single-phase diagram of a three-phase power system with three generators and three loads.

Transient Stability Problem in a Multimachine Network

- Input mechanical power is considered constant.
- Electrical loads are assumed constant and converted to equivalent passive admittances to ground.
- Losses and damping are neglected.

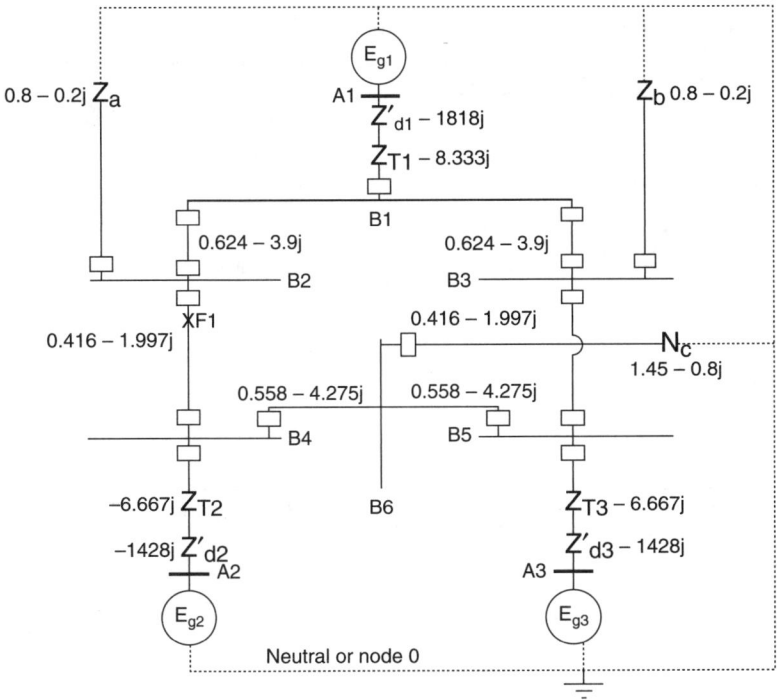

FIGURE 4-2 One-phase admittance diagram including loads, generators, and transformers, 230-kV and 300-MVA base.

Generator	G1	G2	G3
Type	Steam	Combustion turbine	Combustion turbine
MVA	500	300	300
kV	18	18	18
RPM	3600	3600	3600
H $\frac{MW \cdot sec}{MVA}$	6	4	4
X_d	1.2 per unit	0.95 per unit	0.95 per unit
X_d'	0.15 per unit	0.12 per unit	0.12 per unit

TABLE 4-1 Generator Factory Ratings

4.1 Minimum Data Necessary to Do a Transient Stability Study

Besides obtaining a single-line diagram of the entire network showing all the transmission line impedances and all the apparatus connected to the network, you need the latest load flow study available showing the demand of real and reactive power at all the busses. Assume that from the information available you extracted the data listed in Tables 4-1 to 4-6 and in Figs. 4-2 and 4-3.

Let us select 300 MVA and 230 kV as the bases for the entire power system. The new values of the generators' inertial constants are computed using Eq. (4.1):

$$\text{MVA}_{new} \cdot H_{new} = H_{given} \cdot \text{MVA}_{generator} \qquad (4.1)$$

$$H_{new} = H_{given} \frac{\text{MVA}_{generator}}{\text{MVA}_{new}}$$

Generator	G1	G2	G3	G2G3
MVA	1.67	1	1	2
kV	0.078	0.078	0.078	0.078
H $\frac{MW \cdot sec}{MVA}$	10	4	4	8
X_d	0.0044	0.0058	0.0058	0.0029
X_d'	0.00055	0.0007	0.0007	0.00035

TABLE 4-2 Generator Per-Unit Values Converted to 230-kV and 300-MVA Bases

Transformer	T1	T2	T3	T2T3
MVA	1.667	1	1	2
kV	1	1	1	1
Ratio	12.78	12.78	12.78	12.78
X_T	0.12	0.15	0.15	0.075
Y_T	−8.333	−6.667	−6.667	−13.333j

TABLE 4-3 Transformer Per-Unit Values Converted to 230-kV and 300-MVA Bases

Line	R	X
A1–B1		0.0005476
B1–B2	0.04	0.25
B1–B3	0.04	0.25
B2–B4	0.1	0.48
B3–B5	0.1	0.48
B4–B6	0.03	0.23
B5–B6	0.03	0.23
B4–A2		0.00069
B5–A3		0.00069

Base: 230 kV, 300 MVA

TABLE 4-4 Line Impedances, Per-Unit

Line	G	B
A1–B1		−1826
B1–B2	0.624	−3.9
B1–B3	0.624	−3.9
B2–B4	0.416	−1.997
B3–B5	0.416	−1.997
B4–B6	0.558	−4.275
B5–B6	0.558	−4.275
B4–A2		−1435
B5–A3		−1435

Base: 230 kV, 300 MVA

TABLE 4-5 Line Admittances, Per-Unit

Bus	P	Q	V	Name
B1			1.05 arg(0)	Swing bus/generator G1 bus
B2	−0.8	−0.2	1.00 arg(0)	Inductive load bus/constant Q
B3	−0.8	−0.2	1.00 arg(0)	Inductive load bus/constant Q
B4	1.0		1.05 arg(0)	Voltage magnitude constant/G2 bus
B5	1.0		1.05 arg(0)	Voltage magnitude constant/G3 bus
B6	−1.45	−0.8	1.00 arg(0)	Inductive load bus/constant Q

Base: 230 kV, 300 MVA

TABLE 4-6 Estimated Per-Unit Real and Reactive Power Input to Network and Bus Voltage

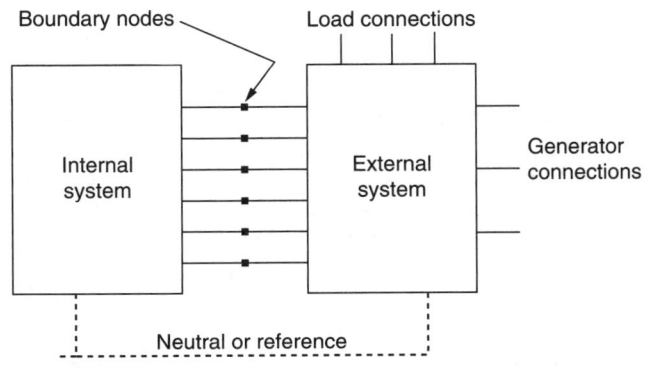

FIGURE 4-3 Power system split to facilitate analysis.

The values of H for G2 and G3 do not change. The new value for G1 is

$$H_{new} := 6 \cdot \frac{500}{300} \qquad H_{new} = 10 \qquad \text{per unit}$$

The new per-unit values of the generator reactances are

$$X_{new} = X_{given} \cdot \left(\frac{kV_{given}}{kV_{new}}\right)^2 \cdot \frac{MVA_{new}}{MVA_{given}} \qquad \text{See Eq. (1.4).}$$

$$\left(\frac{kV_{given}}{kV_{new}}\right)^2 = \left(\frac{18}{230}\right)^2 = 0.006125$$

Therefore,

$$X_{new} = X_{given} \cdot \frac{300}{MVA_{given}} \cdot (6.125 \times 10^{-3}) \qquad (4.2)$$

For G1:

$$X_{new} = X_{given} \cdot \frac{300}{500} (6.125 \times 10^{-3})$$

$1.2 \times 0.6 \times (6.125 \times 10^{-3}) = 0.0044 \qquad X_{dnew} = 0.0044$

$0.15 \times 0.6 \times (6.125 \times 10^{-3}) = 0.00055 \qquad X'_{dnew} = 0.00055$

For G2 and G3:

$$X_{new} = X_{given} \cdot (6.125 \times 10^{-3}) \cdot \frac{300}{300}$$

$0.95 \times (6.125 \times 10^{-3}) = 0.0058 \qquad X_{dnew} = 0.0058$

$0.12 \times (6.125 \times 10^{-3}) = 0.0007 \qquad X'_{dnew} = 0.0007$

Figure 4-2 shows an illustration of the entire power system. However, it is convenient to divide the system into internal and external subsystems as illustrated in Fig. 4-3. The external subsystem contains all the generators, transformers, and loads; and the internal subsystem contains everything else. Both subsystems are connected through the boundary nodes. And it is assumed that any fault would occur inside the internal subsystem.

At the swing bus B1, we specify the voltage magnitude and phase angle that are to be held constant. At the other generator-transformer busses, B4 and B5, we specify the voltage magnitude and hold

Transient Stability Problem in a Multimachine Network

them constant. At load busses B2, B3, and B6, we specify the reactive power flow. The effect of increasing the specified voltage at a generator bus is to produce an increase of the reactive power delivered to the bus by the generator. The internal system data are summarized in Tables 4-4 to 4-6. The admittances and impedances of the power system illustrated in Fig. 4-2 are given below.

$$Z_{1,2} := 0.04 + 0.25j \qquad Y_{1,2} := \frac{1}{Z_{1,2}} 0.624 - 3.9j$$

$$Z_{2,4} := 0.1 + 0.48j \qquad Y_{2,4} := \frac{1}{Z_{2,4}} 0.416 - 1.997j$$

$$Z_{4,6} := 0.03 + 0.23j \qquad Y_{4,6} := \frac{1}{Z_{4,6}} = 0.558 - 4.275j$$

The total three-phase loads connected to busses B2, B3, and B6 are inductive loads. Symbolically,

$$\frac{0.2}{0.8} = 0.25 \qquad \text{atan } 0.25 = 14.036° \qquad \text{inductive load}$$

$$\frac{0.8}{1.45} = 0.552 \qquad \text{atan } 0.552 = 28.899° \qquad \text{inductive load}$$

4.2 Converting Electrical Loads to Equivalent Admittances

The procedure to convert loads A, B, and C to equivalent line-to-neutral admittances is given below. It replaces each load with a constant admittance. However, keep in mind that the load representation used can have a marked effect on stability results, because loads are very dynamic electrical objects that could change their values at any time and that can produce spikes and harmonics. So the equivalent constant admittance is not the best representation of an electrical load. However, constant load assumption is commonly used because in this way the nodes to which they are connected could be eliminated, which greatly facilitates the analysis.

Let us define load A as follows:

V_2 bus B2 line-to-line voltage
I_2 current flowing into load admittance
$Y_A = G_A + B_A \cdot j$ load admittance
$P_2 + Q_2 \cdot j$ load real and reactive power
$P_2 + Q_2 \cdot j = V_2 \cdot \bar{I}_2$ where $\bar{I}_2 = V_2 \cdot (G_A - B_A \cdot j)$ conjugate

$$P_2 + Q_2 \cdot j = V_2 [V_2 \cdot (G_A - B_A \cdot j)] = (|V_2|)^2 (G_A - B_A \cdot j)$$
$$= ([V_2])^2 \cdot G_A - (|V_2|)^2 B_A \cdot j$$

$$G_A = \frac{P_2}{(|V_2|)^2} \qquad B_A = \frac{-Q_2}{(|V_2|)^2}$$

Therefore the equivalent load admittance is:

$$Y_A = \frac{P_2}{(|V_2|)^2} - \frac{Q_2}{(|V_2|)^2} \cdot j \qquad (4.3)$$

Similarly for loads B and C,

$$Y_B = \frac{P_3}{(|V_3|)^2} - \frac{Q_3}{(|V_3|)^2} \cdot j \qquad Y_C = \frac{P_6}{(|V_6|)^2} - \frac{Q_6}{(|V_6|)^2} \cdot j$$

In Table 4-6 the three-phase power delivered to the loads appears as negative input power delivered to the network, but it is positive input power delivered to the load. Because the three-phase power system is assumed symmetrical and balanced, the analysis is performed on a per-phase base; so the power delivered from phase to neutral is only one-third of the total three-phase power. However, for per-unit computational purposes it is the same to use one-third of the power and one-third of the three-phase MVA base as to use the total power and the total three-phase MVA base.

Base MVA = 300

$P_2 = 0.8$ per-unit three-phase real power $P_6 = 1.45$
$Q_2 = 0.2$ per-unit three-phase reactive power $Q_6 = 0.8$
$V_2 = 1.00$ per-unit line-to-line voltage $V_6 = 1.00$

$$Y_A = \frac{0.8}{1} - \frac{0.2}{1} j \qquad Y_A := 0.8 - 0.2j \qquad \text{per-unit equivalent admittance of load A}$$

$$Y_B = \frac{0.8}{1} - \frac{0.2}{1} j \qquad Y_B := 0.8 - 0.2j \qquad \text{per-unit equivalent admittance of load B}$$

$$Y_C = \frac{1.45}{1} - \frac{0.8}{1} j \qquad Y_C := 1.45 - 0.8j \qquad \text{per-unit equivalent admittance of load C}$$

4.3 Load Flow during Normal Operation

Base three-phase MVA = 300 MVA

Base line-to-line voltage = 230 kV

Base admittance: $\dfrac{300}{230^2} = 5.671 \times 10^{-3} = 0.005671$ mho

Transformer's per-unit reactances are listed in Table 4-3, and lines per-unit admittances are given in Fig. 4-2 and listed in Table 4-5. B1 is selected as the swing bus.

The classical model (see Fig. 4-4) is very useful in stability analysis, but is limited to the study of transients for only the first swing or for a period of approximately 1 second (s). This is the period of time during which the system dynamic response depends basically on the stored kinetic energy in the rotating generator-driver masses.

The bus admittance matrix or the nodal admittance matrix of the internal system depicted in Figs. 4-2 and 4-3 is

$$Y_{bus} := \begin{pmatrix} Y_{1,1} & Y_{1,2} & Y_{1,3} & Y_{1,4} & Y_{1,5} & Y_{1,6} \\ Y_{2,1} & Y_{2,2} & Y_{2,3} & Y_{2,4} & Y_{2,5} & Y_{2,6} \\ Y_{3,1} & Y_{3,2} & Y_{3,3} & Y_{3,4} & Y_{3,5} & Y_{3,6} \\ Y_{4,1} & Y_{4,2} & Y_{4,3} & Y_{4,4} & Y_{4,5} & Y_{4,6} \\ Y_{5,1} & Y_{5,2} & Y_{5,3} & Y_{5,4} & Y_{5,5} & Y_{5,6} \\ Y_{6,1} & Y_{6,2} & Y_{6,3} & Y_{6,4} & Y_{6,5} & Y_{6,6} \end{pmatrix}$$

where the self-admittances are identified by repeated subscripts and each node (bus) self-admittance equals the sum of all the admittances terminating on that node.

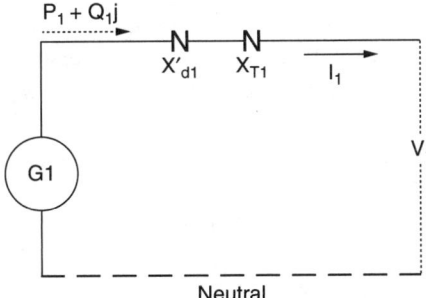

FIGURE 4-4 Classical generator and transformer combination model.

The rest of the matrix's admittances are the mutual admittances between nodes, and each equals the negative of the sum of all admittances connected directly between the nodes identified by the double subscript.

The bus admittance matrix is a square matrix usually designated as Y_{bus}. For a power system with N busses, the admittance matrix is a square N × N symmetric matrix in which symmetrically located off-diagonal elements are equal, for instance, $Y_{1,2} = Y_{2,1}$.

The diagonal elements are the self-admittance terms, and the off-diagonal elements are the mutual admittance between nodes. The bus admittance matrix of a large power system usually contains many zero elements, and it is classified as a *sparse* matrix. The bus admittance matrix relates all the bus current injections to the bus voltages. In fact, it is the matrix version of Ohm's law. Symbolically,

$$I = Y_{bus} V$$

The internal system depicted in Figs. 4-2 and 4-3 excluding generators, transformers, and loads has the following six independent node equations.

$$Y_{1,1} \cdot V_1 + Y_{1,2} \cdot V_2 + Y_{1,3} \cdot V_3 + Y_{1,4} \cdot V_4 + Y_{1,5} \cdot V_5 + Y_{1,6} \cdot V_6 = I_1$$
$$Y_{2,1} \cdot V_1 + Y_{2,2} \cdot V_2 + Y_{2,3} \cdot V_3 + Y_{2,4} \cdot V_4 + Y_{2,5} \cdot V_5 + Y_{2,6} \cdot V_6 = I_2$$
$$Y_{3,1} \cdot V_1 + Y_{3,2} \cdot V_2 + Y_{3,3} \cdot V_3 + Y_{3,4} \cdot V_4 + Y_{3,5} \cdot V_5 + Y_{3,6} \cdot V_6 = I_3$$
$$Y_{4,1} \cdot V_1 + Y_{4,2} \cdot V_2 + Y_{4,3} \cdot V_3 + Y_{4,4} \cdot V_4 + Y_{4,5} \cdot V_5 + Y_{4,6} \cdot V_6 = I_4$$
$$Y_{5,1} \cdot V_1 + Y_{5,2} \cdot V_2 + Y_{5,3} \cdot V_3 + Y_{5,4} \cdot V_4 + Y_{5,5} \cdot V_5 + Y_{5,6} \cdot V_6 = I_5$$
$$Y_{6,1} \cdot V_1 + Y_{6,2} \cdot V_2 + Y_{6,3} \cdot V_3 + Y_{6,4} \cdot V_4 + Y_{6,5} \cdot V_5 + Y_{6,6} \cdot V_6 = I_6$$

Node voltages are unknown. If the voltage at any node is known in magnitude and angle, we do not need to write the equation for that node. The neutral is the usual reference node of the power system.

We will use the Gauss-Seidel method to solve the above system of simultaneous linear equations. In matrix format they could be expressed as $[Y_{bus}] [V] = [I]$, where $[Y_{bus}]$ is the coefficient matrix, $[V]$ is the solution vector, and $[I]$ is the right-hand vector.

$$\begin{pmatrix} Y_{1,1} & Y_{1,2} & Y_{1,3} & Y_{1,4} & Y_{1,5} & Y_{1,6} \\ Y_{2,1} & Y_{2,2} & Y_{2,3} & Y_{2,4} & Y_{2,5} & Y_{2,6} \\ Y_{3,1} & Y_{3,2} & Y_{3,3} & Y_{3,4} & Y_{3,5} & Y_{3,6} \\ Y_{4,1} & Y_{4,2} & Y_{4,3} & Y_{4,4} & Y_{4,5} & Y_{4,6} \\ Y_{5,1} & Y_{5,2} & Y_{5,3} & Y_{5,4} & Y_{5,5} & Y_{5,6} \\ Y_{6,1} & Y_{6,2} & Y_{6,3} & Y_{6,4} & Y_{6,5} & Y_{6,6} \end{pmatrix} \cdot \begin{pmatrix} V_1 \\ V_2 \\ V_3 \\ V_4 \\ V_5 \\ V_6 \end{pmatrix} = \begin{pmatrix} I_1 \\ I_2 \\ I_3 \\ I_4 \\ I_5 \\ I_6 \end{pmatrix} \quad \text{or} \quad Y \cdot V = I$$

(4.4)

Transient Stability Problem in a Multimachine Network

Before we go ahead with the implementation of the Gauss-Seidel (G-S) iterative procedure, we need to determine if the set of solutions generated by the G-S iterative procedure converges to the exact solution value, one for each variable. Solution convergence occurs if the [Y] coefficient matrix is diagonally dominant; if not, the labor-intensive and time-consuming iterations may or may not converge.

The [Y] matrix is diagonally dominant if the magnitude of each diagonal element is greater than or equal to the sum of the other elements' magnitudes in the same row.

$$|Y_{k,k}| \geq \left(\sum_{n=1}^{6} |Y_{k,n}| \right)_{k \neq n} \quad (4.5)$$

or if for at least one diagonal element its magnitude is greater than the sum of the magnitudes of the other elements in the same row.

$$|Y_{k,k}| > \left(\sum_{n=1}^{6} |Y_{k,n}| \right)_{k \neq n} \quad (4.6)$$

Disregarding generators, transformers, and load admittances because they are considered external to the network, we obtain the network admittances from Fig. 4-2 and Table 4-5.

$Y_{1,1} := 1.248 - 7.8j$ $Y_{2,1} := -0.624 + 3.9 \cdot j$ $Y_{3,1} := -0.624 + 3.9 \cdot j$

$Y_{1,2} := -0.624 - 3.9j$ $Y_{2,2} := 1.04 - 5.897j$ $Y_{3,2} := 0$

$Y_{1,3} := -0.624 + 3.9j$ $Y_{2,3} := 0$ $Y_{3,3} := 1.04 - 5.897j$

$Y_{1,4} := 0$ $Y_{2,4} := -0.416 + 1.997j$ $Y_{3,4} := 0$

$Y_{1,5} := 0$ $Y_{2,5} := 0$ $Y_{3,5} := -0.416 + 1.997j$

$Y_{1,6} := 0$ $Y_{2,6} := 0$ $Y_{3,6} := 0$

$Y_{4,1} := 0$ $Y_{5,1} := 0$ $Y_{6,1} := 0$

$Y_{4,2} := -0.416 + 1.997j$ $Y_{5,2} := 0$ $Y_{6,2} := 0$

$Y_{4,3} := 0$ $Y_{5,3} := -0.416 + 1.997j$ $Y_{6,3} := 0$

$Y_{4,4} := 0.974 - 6.272j$ $\quad$ $Y_{5,4} := 0$ $\quad\quad\quad\quad$ $Y_{6,4} := -0.558 + 4.275j$

$Y_{4,5} := 0$ $\quad\quad\quad\quad\quad$ $Y_{5,5} := 0.974 - 6.272j$ $\quad$ $Y_{6,5} := -0.558 + 4.275j$

$Y_{4,6} := -0.558 + 4.275j$ $\quad$ $Y_{5,6} := -0.558 + 4.275j$ $\quad$ $Y_{6,6} := 1.116 - 8.55j$

The bus admittance matrix [Y] of the pretransient original network is written below.

$$Y := \begin{pmatrix} 1.248 - 7.8j & -0.624 + 3.9j & -0.624 + 3.9j & 0 & 0 & 0 \\ -0.624 + 3.9j & 1.04 - 5.897j & 0 & -0.416 + 1.997j & 0 & 0 \\ -0.624 + 3.9j & 0 & 1.04 - 5.897j & 0 & -0.416 + 1.997j & 0 \\ 0 & -0.416 + 1.997j & 0 & 0.974 - 6.272j & 0 & -0.558 + 4.275j \\ 0 & 0 & -0.416 + 1.997j & 0 & 0.974 - 6.272j & -0.558 + 4.275j \\ 0 & 0 & 0 & -0.558 + 4.275j & -0.558 + 4.275j & 1.116 - 8.55j \end{pmatrix}$$

In the square bus admittance matrix, the algebraic sum of all the elements in any row is equal to the shunt admittance connecting the specific row (bus) to ground, or neutral or node zero. Shunt terms in power system models, such as in the π line model of a transmission line, only impact the matrix diagonal elements.

$$Z_{bus} = \frac{1}{Y_{bus}} = Y_{bus}^{-1} \quad\quad Z_{bus} \neq Z$$

If the impedances to ground of all the phases of the power system are infinity, which implies that there is no path to ground, such as the transmission line capacitance to ground, then the sum of the elements in each row of the square bus admittance matrix will be zero, its determinant will also be zero, and therefore the inverse of the bus admittance matrix will not exist. This implies that the bus impedance matrix will not exist either. The matrix Y_{bus} could be directly obtained from circuit diagrams or data. And its inverse is designated as Z_{bus}. However, the matrix Z formed with the circuit impedances is different from the matrix Z_{bus}.

Gauss-Seidel convergence criterion:

The bus admittance matrix is diagonally dominant if it complies with either Eqs. (4.5) or (4.6).

$$|1.248 - 7.8 \cdot j| = 7.89920907$$

$$2 \cdot |-0.624 + 3.9j| = 7.89920907 \quad \text{OK, they are equal}$$

Transient Stability Problem in a Multimachine Network

$|1.04 - 5.897 \cdot j| = 5.988$

$|-0.624 + 3.9 \cdot j| + |-0.416 + 1.997 \cdot j| = 5.989$ No, this is larger

$|1.04 - 5.897j| = 5.988$

$|-0.624 + 3.9j| + |-0.416 + 1.997 \cdot j| = 5.989$ No, this is larger

$|0.974 - 6.272j| = 6.347$

$|-0.416 + 1.997j| + |-0.558 + 4.275j| = 6.351$ No, this is larger

$|0.974 - 6.272j| = 6.347$

$|-0.416 + 1.997j| + |-0.558 + 4.275j| = 6.351$ No, this is larger

$|1.116 + 8.55j| = 8.62252608$

$2 \cdot |-0.558 + 4.275j| = 8.62252608$ OK, they are equal

There are only two diagonal terms whose magnitudes are equal to the sum of the magnitudes of the other terms in the same row, and none is larger. So the bus matrix is not diagonally dominant and convergence is not secured. However, it still may converge. Let us try anyway.

Gauss-Seidel Method

Select B1 as the swing bus, start the Gauss-Seidel procedure from bus B2, and do not work with the matrix but with one individual node at a time, for instance, node 2.

Equation for node 2:

$$Y_{2,1} \cdot V_1 + Y_{2,2} \cdot V_2 + Y_{2,3} \cdot V_3 + Y_{2,4} \cdot V_4 + Y_{2,5} \cdot V_5 + Y_{2,6} \cdot V_6 = I_2$$

The power entering the system at bus 2 is

$$V_2 \cdot \overline{I}_2 = P_2 + Q_2 j \quad \text{or} \quad \overline{V}_2 I_2 = P_2 - Q_2 j$$

$$I_2 = \frac{P_2 - Q_2 j}{\overline{V}_2}$$

Therefore, the equation for node 2 becomes

$$Y_{2,1} \cdot V_1 + Y_{2,2} \cdot V_2 + Y_{2,3} \cdot V_3 + Y_{2,4} \cdot V_4 + Y_{2,5} \cdot V_5 + Y_{2,6} \cdot V_6 = \frac{P_2 - Q_2 \cdot j}{\overline{V_2}}$$

$$Y_{2,2} \cdot V_2 = \frac{P_2 - Q_2 \cdot j}{\overline{V_2}} - (Y_{2,1} \cdot V_1 + Y_{2,3} \cdot V_3 + Y_{2,4} \cdot V_4 + Y_{2,5} \cdot V_5 + Y_{2,6} \cdot V_6)$$

$$V_2 = \frac{1}{Y_{2,2}} \left[\frac{P_2 - Q_2 \cdot j}{\overline{V_2}} - (Y_{2,1} \cdot V_1 + Y_{2,3} \cdot V_3 + Y_{2,4} \cdot V_4 + Y_{2,5} \cdot V_5 + Y_{2,6} \cdot V_6) \right]$$

(4.7)

Generalizing,

$$V_k = \frac{1}{Y_{k,k}} \left[\frac{P_k - Q_k \cdot j}{\overline{V_k}} - \sum_{n=1}^{N} (Y_{k,n} \cdot V_n) \right] \quad n \neq k,\ N = \text{number of busses}$$

(4.8)

Table 4-6 provides the estimated values of the magnitude of the bus voltages and the real and reactive power demanded at the load busses. Equation (4.7) provides the magnitude and angle of the bus voltage with respect to the bus input current. Usually the neutral is selected as the reference node or node zero for the entire power system. The power input into bus B2 is listed as negative in Table 4-6, because it is power leaving the bus and delivered to the load.

$V_1 := 1.05 \quad V_2 := 1.00 \quad V_3 := 1.00 \quad V_4 := 1.05 \quad V_5 := 1.05 \quad V_6 := 1.00$
$\overline{V}_1 = 1.05 \quad \overline{V}_2 = 1 \quad \overline{V}_3 = 1.00 \quad \overline{V}_4 = 1.05 \quad \overline{V}_5 = 1.05 \quad \overline{V}_6 = 1$
$P_2 := -0.8 \quad Q_2 := -0.2 \quad P_3 := -0.8 \quad Q_3 := -0.2 \quad P_6 := -1.45 \quad V_6 := -0.8$

First Iteration
Start with the load busses because the real and reactive powers were previously estimated. Therefore, we can obtain V_2 by solving Eq. (4.7), keeping in mind that in each row of the bus admittance matrix there are three zero elements. For instance, $Y_{2,3}$, $Y_{2,5}$, and $Y_{2,6}$ are all equal to zero.

Computation of V_2:

$$\frac{1}{Y_{2,2}} \left[\frac{-0.8 + 0.2j}{\overline{V_2}} - (Y_{2,1} \cdot V_1 + Y_{2,4} \cdot V_4) \right] = 0.994 - 0.126j \qquad (4.9)$$

$$V_2 := 0.994 - 0.126j \qquad \overline{V}_2 := 0.994 + 0.126j \qquad |V_2| = 1.002$$

$$\frac{1}{Y_{2,2}} \cdot \left[\frac{-0.8 + 0.2j}{\overline{V}_2} - (Y_{2,1} V_1 + Y_{2,4} \cdot V_4) \right] = 0.979 - 0.117j$$

We have including new conjugate

$$V_2 := 0.979 - 0.117j \qquad \overline{V}_2 := 0.979 + 0.117j \qquad |V_2| = 0.986 \qquad \text{includes new conjugate}$$

Computation of V_3:

$$\frac{1}{Y_{3,3}} \left[\frac{-0.8 + 0.2j}{\overline{V}_3} - (Y_{3,1} \cdot V_1 + Y_{3,5} \cdot V_5) \right] = 0.994 - 0.126j$$

$$V_3 := 0.994 - 0.126j \qquad \overline{V}_3 = 0.994 + 0.126j \qquad |V_3| = 1.002$$

$$\frac{1}{Y_{3,3}} \left[\frac{-0.8 + 0.2j}{\overline{V}_3} - (Y_{3,1} \cdot V_1 + Y_{3,5} \cdot V_5) \right] = 0.979 - 0.117j$$

$$V_3 := 0.979 - 0.117j \qquad \overline{V}_3 := 0.979 + 0.117j \qquad |V_3| = 0.986 \qquad \text{includes new conjugate}$$

Computation of V_6:

$$\frac{1}{Y_{6,6}} \left[\frac{-1.45 + 0.8j}{\overline{V}_6} - (Y_{6,1} \cdot V_1 + Y_{6,2} \cdot V_2 + Y_{6,3} \cdot V_3 + Y_{6,4} \cdot V_4 + Y_{6,5} \cdot V_5) \right]$$
$$= 0.9362 - 0.1547j$$

$$V_6 := 0.9362 - 0.1547j \qquad \overline{V}_6 = 0.936 + 0.155j \qquad |V_6| = 0.949$$

$$\frac{1}{Y_{6,6}} \left[\frac{-1.45 + 0.8j}{\overline{V}_6} - (Y_{6,1} \cdot V_1 + Y_{6,2} \cdot V_2 + Y_{6,3} \cdot V_3 + Y_{6,4} \cdot V_4 + Y_{6,5} \cdot V_5) \right]$$
$$= 0.905 - 0.141j$$

$$V_6 := 0.905 - 0.141j \qquad \overline{V}_6 := 0.905 + 0.141j \qquad |V_6| = 0.916 \qquad \text{includes new conjugate}$$

Equations (4.7) and (4.8) apply only to busses in which the real and reactive powers are specified. Since B1, B4, and B5 are generator busses and the voltage magnitude is specified instead of the reactive power, first we compute the value of the input reactive power at B4, and for the real power input we use the value estimated in Table 4-6. Then we compute the real and imaginary components of the bus voltage at B4. And we repeat the procedure for B5.

From Eq. (4.8) we obtain

$$Y_{k,k} \cdot V_k = \frac{P_k - Q_k \cdot j}{\overline{V}_K} - \left[\sum_{n=1}^{N}(Y_{k,n} \cdot V_n)\right] \quad n \neq k$$

$$\frac{P_k - Q_k \cdot j}{\overline{V}_K} = Y_{kk} \cdot V_k + \left[\sum_{n=1}^{N}(Y_{k,n} \cdot V_n)\right] \quad n \neq k$$

Allowing n to be equal to k, we arrive at

$$\frac{P_k - Q_k j}{\overline{V}_K} = \sum_{n=1}^{N}(Y_{k,n} \cdot V_n)$$

$$P_k - Q_k j = \overline{V}_k \cdot \sum_{n=1}^{N}(Y_{k,n} \cdot V_n) \quad \text{n is allowed to equal k} \quad (4.10)$$

The B4 reactive power computation for k := 1,2.. 6 and N := 6

$$P_4 - Q_k j = \overline{V}_4 \sum_{n=1}^{N}(Y_{4,n} \cdot V_n) \quad \text{per unit}$$

The imaginary part of the power complex number is

$$-\text{Im}\left[\overline{V}_4 \cdot \sum_{n=1}^{N}(Y_{4,n} \cdot V_n)\right] = 0.666 \quad Q_4 := 0.666 \quad P_4 := 1$$

Substituting in Eq. (4.8), we obtain

$$V_4 = \frac{1}{Y_{4,4}} \cdot \left[\frac{P_4 - Q_4 j}{\overline{V}_4} - (Y_{4,1} \cdot V_1 + Y_{4,2} \cdot V_2 + Y_{4,3} \cdot V_3 + Y_{4,5} \cdot V_5 + Y_{4,6} \cdot V_6)\right]$$

$$V_4 := 1.050129 + 0.000859j \quad |V_4| = 1.050129$$

This value of complex V_4 must be corrected because $|V_4|$ is not 1.05 as specified in Table 4-6. The corrected complex V_4 must have a 1.05 magnitude. So the corrected complex V_4 is

$$V_4 := \frac{1.05}{1.050129} \cdot (1.050129 + 0.000859j) \quad V_4 = 1.05 + 0.001j \quad |V_4| = 1.05$$

$$V_4 := 1.05 + 0.001j \quad \text{corrected value of complex } V_4$$

Transient Stability Problem in a Multimachine Network

Computation of V_5:

$$Q_5 := -\operatorname{Im}\left[\overline{V}_5 \sum_{n=1}^{N}(Y_{5,n} \cdot V_n)\right] = 0.666 \qquad P_5 := 1$$

$$V_5 = \frac{1}{Y_{5,5}} \cdot \left[\frac{P_5 - Q_5 j}{\overline{V}_5} - (Y_{5,1} \cdot V_1 + Y_{5,2} \cdot V_2 + Y_{5,3} \cdot V_3 + Y_{5,4} \cdot V_4 + Y_{5,6} \cdot V_6)\right]$$

$V_5 := 1.050133 + 0.000858j \qquad |V_5| = 1.050133$

Correction:

$$V_5 := \frac{1.05}{1.050133} \cdot (1.050133 + 0.000858j) \quad V_5 = 1.05 + 0.001j \quad |V_5| = 1.05$$

$V_5 := 1.05 + 0.001j \qquad$ corrected value of complex V_5

Computation of V_1:
Substituting in Eq. (4.10), we obtain Q_1:

$$P_1 - Q_1 j = \overline{V}_1 \sum_{n=1}^{6}(Y_{1,n} \cdot V_n) \qquad Q_1 := -\operatorname{Im}\left[\overline{V}_1 \cdot \sum_{n=1}^{6}(Y_{1,n} \cdot V_n)\right] = 0.428$$

Substituting in Eq. (4.8), we obtain V_1:

$$V_1 = \frac{1}{Y_{1,1}}\left[\frac{P_1 - Q_1 j}{\overline{V}_1} - (Y_{1,2} \cdot V_2 + Y_{1,3} \cdot V_3 + Y_{1,4} \cdot V_4 + Y_{1,5} V_5 + Y_{1,6} \cdot V_6)\right]$$

$= 1.029975 - 0.125156j$

$V_1 := 1.029975 - 0.125156j \qquad \overline{V}_1 = 1.03 + 0.1252j \qquad |V_1| = 1.037551$

Swing bus voltage magnitude correction:

$$V_1 := \frac{1.05}{1.037551}(1.029975 - 0.125156j) = 1.042 - 0.127j$$

$V_1 := 1.042 - 0.127j \qquad \overline{V}_1 = 1.042 + 0.127j \qquad |V_1| = 1.05$

Set of bus voltages before completion of the first iteration

$V_1 := 1.042 - 0.127j2 \qquad \overline{V}_1 = 1.042 + 0.127j \qquad |V_1| = 1.05$
$V_2 := 0.979 - 0.117j \qquad \overline{V}_2 = 0.979 + 0.117j \qquad |V_2| = 0.986$
$V_3 := 0.979 - 0.117j \qquad \overline{V}_3 = 0.979 + 0.117j \qquad |V_3| = 0.986$

$V_4 := 1.05 + 0.001j$ $\overline{V}_4 = 1.05 - 0.001j$ $|V_4| = 1.05$

$V_5 := 1.05 + 0.001j$ $\overline{V}_5 = 1.05 - 0.001j$ $|V_5| = 1.05$

$V_6 := 0.905 - 0.141j$ $\overline{V}_6 = 0.905 + 0.141j$ $|V_6| = 0.916$

Set of real and reactive powers before completion of the first iteration:

$P_1 := 1$ $Q_1 :=$ $P_4 := 1$ $Q_4 := 0.666$

$P_2 := -0.8$ $Q_2 := -0.2$ $P_5 := 1$ $Q_5 := 0.666$

$P_3 := -0.8$ $Q_3 := -0.2$ $P_6 := -1.45$ $Q_6 := -0.8$

Balance of power computations:

Power into network = Power out of network

Generators supply power to network

Loads receive power out of network

$P_1 + (P_4 + P_5) = P_1 + 2$ $-(P_2 + P_3 + P_6) = 3.05$

$P_1 + 2 = 3.05$ $P_1 := 1.05$

$Q_1 + (Q_4 + Q_5) = Q_1 + 1.332$ $-(Q_2 + Q_3 + Q_6) = 1.2$

$Q_1 + 1.332 = 1.2$ $Q_1 := -0.132$

Real power balance:

$$(P_1 + P_4 + P_5) + (P_2 + P_3 + P_6) = 0$$

Reactive power balance:

$$(Q_1 + Q_4 + Q_5) + (Q_2 + Q_3 + Q_6) = 0$$

Generators G2 and G3, connected to buses B4 and B5 respectively, are making available to the load too much reactive power, 1.332 MVAR, when the load only needs 1.2 MVAR. The swing bus B1 needs to take the surplus or 0.132 MVAR. Instead of using the value of Q_1 computed using Eq. (4.10), we choose the value provided by the balance of power equations for Q_1 and P_1 from them we could find V_1 as follows:

$$V_1 := \frac{1}{Y_{1,1}} \left[\frac{P_1 - Q_1 j}{\overline{V}_1} - (Y_{1,2} \cdot V_2 + Y_{1,3} \cdot V_3 + Y_{1,4} \cdot V_4 + Y_{1,5} \cdot V_5 + Y_{1,6} \cdot V_6) \right]$$

$= 0.999 + 0.009j$

$V_1 = 0.999 + 0.009j$ $\overline{V}_1 = 0.999 + 0.009j$ $|V_1| = 0.999$

Transient Stability Problem in a Multimachine Network 83

Swing bus voltage magnitude correction:

$$V_1 := \frac{1.05}{0.999} \cdot (0.999 + 0.009j) \qquad V_1 = 1.05 + 0.009j$$

$$V_1 := 1.05 + 0.009j \qquad \overline{V}_1 = 1.05 - 0.009j \qquad |V_1| = 1.05$$

After the first iteration we have

$$V_1 := 1.05 + 0.009j \quad \overline{V}_1 = 1.05 - 0.009j \quad |V_1| = 1.05 \quad \arg(V_1) = 0.491°$$
$$V_2 := 0.979 - 0.117j \quad \overline{V}_2 = 0.979 + 0.117j \quad |V_2| = 0.986 \quad \arg(V_2) = -6.815°$$
$$V_3 := 0.979 - 0.117j \quad \overline{V}_3 = 0.979 + 0.117j \quad |V_3| = 0.986 \quad \arg(V_3) = -6.815°$$
$$V_4 := 1.05 + 0.001j \quad \overline{V}_4 = 1.05 - 0.001j \quad |V_4| = 1.05 \quad \arg(V_4) = 0.055°$$
$$V_5 := 1.05 + 0.001j \quad \overline{V}_5 = 1.05 - 0.001j \quad |V_5| = 1.05 \quad \arg(V_5) = 0.055°$$
$$V_6 := 0.905 - 0.141j \quad \overline{V}_6 = 0.905 + 0.141j \quad |V_6| = 0.916 \quad \arg(V_6) = -8.856°$$

The angles shown are the voltage angles with reference to the current. Set of real and reactive powers after the first iteration:

$$\begin{array}{llll} P_1 := 1.05 & Q_1 := -0.132 & P_4 := 1 & Q_4 := 0.666 \\ P_2 := -0.8 & Q_2 := -0.2 & P_5 := 1 & Q_5 := 0.666 \\ P_3 := -0.8 & Q_3 := -0.2 & P_6 := -1.45 & Q_6 := -0.8 \end{array}$$

Let us now compute the absolute relative approximate error, using the following formula for $\varepsilon \equiv$ error and $k := 1..6$.

$$(V_k)^{new} = \begin{pmatrix} 1.05 \\ 0.986 \\ 0.986 \\ 1.05 \\ 1.05 \\ 0.916 \end{pmatrix} \qquad (V_k)^{old} = \begin{pmatrix} 1.05 \\ 1.00 \\ 1.00 \\ 1.05 \\ 1.05 \\ 1.00 \end{pmatrix} \qquad (|\varepsilon|)_k = \left| \frac{(V_k)^{new} - (V_k)^{old}}{(V_k)^{new}} \right| \cdot 100$$

(4.11)

$$\varepsilon_1 := \left|\frac{1.05 - 1.05}{1.05}\right| \cdot 100 = 0 \qquad \varepsilon_3 := \left|\frac{0.986 - 1.00}{0.986}\right| \cdot 100 = 1.42$$

$$\varepsilon_2 := \left|\frac{0.986 - 1.00}{0.986}\right| \cdot 100 = 1.42 \qquad \varepsilon_4 := \left|\frac{1.05 - 1.05}{1.05}\right| \cdot 100 = 0$$

$$\varepsilon_5 := \left|\frac{1.05 - 1.05}{1.05}\right| \cdot 100 = 0$$

$$\varepsilon_6 := \left|\frac{0.916 - 1.00}{0.916}\right| \cdot 100 = 9.17$$

84　Chapter Four

The maximum relative approximate error is 9.17 percent, which is not too bad, but let us try to reduce it to 1 percent or less. Equation (4.8) is applicable to all iterations and is repeated below as a reminder. For $k := 1, 2..\ 6$ and $N = 6$,

$$V_k = \frac{1}{Y_{k,k}}\left[\frac{P_k - Q_k \cdot j}{\overline{V_k}} - \sum_{n=1}^{N}(Y_{k,n} \cdot V_n)\right] \quad n \neq k$$

$$V_k = \begin{pmatrix} 1.05 + 0.009j \\ 0.979 - 0.117j \\ 0.979 - 0.117j \\ 1.05 + 0.001j \\ 1.05 + 0.001j \\ 0.905 - 0.141j \end{pmatrix} \quad |V_k| = \begin{pmatrix} 1.05 \\ 0.986 \\ 0.986 \\ 1.05 \\ 1.05 \\ 0.916 \end{pmatrix} \quad \arg(V_k) = \begin{pmatrix} 0.491 \\ -6.815 \\ -6.815 \\ 0.055 \\ 0.055 \\ -8.856 \end{pmatrix} \text{deg}$$

Second Iteration
Computation of V_2:

$$\frac{1}{Y_{2,2}}\left[\frac{P_2 - Q_2 j}{\overline{V_2}} - (Y_{2,1} \cdot V_1 + Y_{2,4} \cdot V_4)\right] = 0.978 - 0.114j$$

$$V_2 := 0.978 - 0.114j \quad \overline{V_2} = 0.978 + 0.114j$$

$$|V_2| = 0.985 \quad \arg(V_2) = -6.649°$$

Computation of V_3:

$$\frac{1}{Y_{3,3}}\left[\frac{P_3 - Q_3 j}{\overline{V_3}} - (Y_{3,1} \cdot V_1 + Y_{3,5} \cdot V_5)\right] = 0.978 - 0.114j$$

$$V_3 := 0.978 - 0.114j \quad \overline{V_3} = 0.978 + 0.114j$$

$$|V_3| = 0.985 \quad \arg(V_3) = -6.649°$$

Computation of V_6:

$$\frac{1}{Y_{6,6}}\left[\frac{P_6 - Q_6 j}{\overline{V_6}} - (Y_{6,1} \cdot V_1 + Y_{6,2} \cdot V_2 + Y_{6,3} \cdot V_3 + Y_{6,4} \cdot V_4 + Y_{6,5} \cdot V_5)\right]$$

$$= 0.9013 - 0.1468j$$

$$V_6 := 0.9013 - 0.1468j \quad \overline{V_6} = 0.901 + 0.147j \quad |V_6| = 0.913 \quad \arg(V_6) = -9.251°$$

Computation of V_4:

$$\frac{1}{Y_{4,4}} \cdot \left[\frac{P_4 - Q_4 j}{\overline{V_4}} - (Y_{4,1} \cdot V_1 + Y_{4,2} \cdot V_2 + Y_{4,3} \cdot V_3 + Y_{4,5} \cdot V_5 + Y_{4,6} \cdot V_6)\right] = 1.047 - 0.002j$$

$V_4 := 1.047 - 0.002j$ $\quad \overline{V}_4 = 1.047 + 0.002j \quad |V_4| = 1.047002$

This value of complex V_4 must be corrected because $|V_4|$ is not 1.05 as specified in Table 4-6. The corrected complex V_4 must have a 1.05 magnitude. So the corrected complex V_4 is

$$\frac{1.05}{1.047002} \cdot (1.047 - 0.002j) = 1.05 - 0.002j$$

$V_4 := 1.05 - 0.002j \quad \overline{V}_4 = 1.05 + 0.002j \quad |V_4| = 1.05 \quad \arg(V_4) = -0.109°$

Computation of V_5:

$$\frac{1}{Y_{5,5}} \cdot \left[\frac{P_5 - Q_5 j}{\overline{V_5}} - (Y_{5,1} \cdot V_1 + Y_{5,2} \cdot V_2 + Y_{5,3} \cdot V_3 + Y_{5,4} \cdot V_4 + Y_{5,6} \cdot V_6)\right] = 1.047 - 0.002j$$

$V_5 := 1.047 - 0.002j \quad \overline{V}_5 = 1.047 + 0.002j \quad |V_5| = 1.047002$

$V_5 := \frac{1.05}{1.047002} \cdot (1.047 - 0.002j) = 1.05 - 0.002j \quad |V_5| = 1.05 \quad \arg(V_5) = -0.109°$

$V_5 := 1.05 - 0.002j \quad$ corrected value of complex V_5

Computation of V_1:

$$\frac{1}{Y_{1,1}} \cdot \left[\frac{P_1 - Q_1 j}{\overline{V_1}} - (Y_{1,2} \cdot V_2 + Y_{1,3} \cdot V_3 + Y_{1,4} \cdot V_4 + Y_{1,5} V_5 + Y_{1,6} \cdot V_6)\right]$$

$V_1 = 1.05 + 0.009j \quad \overline{V}_1 = 1.05 - 0.009j \quad |V_1| = 1.050039$

Voltage magnitude correction:

$V_1 := \frac{1.05}{1.050039} \cdot (1.05 + 0.009j) = 1.05 + 0.009j \quad$ no correction required

$V_1 := 1.05 + 0.009j \quad \overline{V}_1 = 1.05 - 0.009j \quad |V_1| = 1.05 \quad \arg(V_1) = 0.491°$

The bus voltages in matrix format are given next:

$$V_k = \begin{pmatrix} 1.05+0.009j \\ 0.978-0.114j \\ 0.978-0.114j \\ 1.05-0.002j \\ 1.05-0.002j \\ 0.901-0.147j \end{pmatrix} \quad |V_k| = \begin{pmatrix} 1.05 \\ 0.985 \\ 0.985 \\ 1.05 \\ 1.05 \\ 0.913 \end{pmatrix} \quad \arg(V_k) = \begin{pmatrix} 0.491 \\ -6.649 \\ -6.649 \\ -0.109 \\ -0.109 \\ -9.251 \end{pmatrix} \text{deg}$$

Let us now compute the absolute relative approximate error after the second iteration. For $\varepsilon \equiv$ error and $k := 1 .. 6$,

$$(|V_k|)^{\text{new}} = \begin{pmatrix} 1.05 \\ 0.985 \\ 0.985 \\ 1.05 \\ 1.05 \\ 0.913 \end{pmatrix} \quad (|V_k|)^{\text{old}} = \begin{pmatrix} 1.05 \\ 0.986 \\ 0.986 \\ 1.05 \\ 1.05 \\ 0.916 \end{pmatrix}$$

$$(|\varepsilon|)_k = \left| \frac{(V_k)^{\text{new}} - (V_k)^{\text{old}}}{(V_k)^{\text{new}}} \right| \cdot 100 \qquad (4.12)$$

$$\varepsilon_1 := \left| \frac{1.05 - 1.05}{1.05} \right| \cdot 100 = 0 \qquad \varepsilon_3 := \left| \frac{0.985 - 0.986}{0.985} \right| \cdot 100 = 0.102$$

$$\varepsilon_2 := \left| \frac{0.985 - 0.986}{0.985} \right| \cdot 100 = 0.102 \qquad \varepsilon_4 := \left| \frac{1.05 - 1.05}{1.05} \right| \cdot 100 = 0$$

$$\varepsilon_5 := \left| \frac{1.05 - 1.05}{1.05} \right| \cdot 100 = 0$$

$$\varepsilon_6 := \left| \frac{0.913 - 0.916}{0.913} \right| \cdot 100 = 0.329$$

The maximum relative approximate error has been reduced to 0.329 percent, which is acceptable. At this point we stop the procedure because the intent of this demonstration is to show the implementation of the method and not to arrive to an exact solution. In real cases the number of iterations could be on the order of 50 or more, and they are accomplished by using computer programs. The angles shown are the voltage angles with reference to the current.

4.4 Initial Power Angle Computation

The computations below are pretransient and are based on complex power at the generator nodes or busses.

G1-T1:

$$S_1 = P_1 + Q_1 j$$

$V_1 := 1.05 + 0.009j \quad \arg(V_1) = 0.491° \quad |V_1| = 1.05 \quad$ referenced to neutral

The power entering the system at bus B1 is

$$S_1 := 1.05 - 0.132j \tag{4.13}$$

$\overline{S}_1 := 1.05 + 0.132j \quad I_1 := \dfrac{\overline{S}_1}{\overline{V}_1} = 0.999 + 0.134j \quad |I_1| = 1.008$

$\arg(I_1) = 0.13363 \text{ rad} = 7.656°$

I_1 leads bus voltage V_1 by $7.656 - 0.491 = 7.165 \quad \theta_1 := 7.165 \text{ deg}$

The pretransient emf generated by G1 during steady-state operation is

$$E_{g1} = V_1 + (X'_{d1})j \times I_1$$

$X_{d1}' = 0.00055 + 0.12 \quad$ generator plus transformer reactance

$E_{g1} = 1.05 + 0.009j + (0.00055 + 0.12j) \times (0.999 + 0.134j) \rightarrow$

$E_{g1} = 1.0338 + 0.1294j$

$E_{g1} := 1.0338 + 0.1294j \quad |E_{g1}| = 1.042 \quad \arg(E_{g1}) = 7.135°$

referenced to neutral

The G1-T1 initial power angle is $\arg(E_{g1}) - \arg(1.05 + 0.009j) = 6.64347°$

G1 Initial Power Angle

$$\delta_{01} := 6.643°$$

The emf generated by G1 leads bus B1 voltage V1 by 6.643°. See the phasor diagram in Fig. 4-5.

G2-T2:
The power entering the system at bus B4 is:

$$S_4 := P_4 + Q_4 j = 1 + 0.666j \tag{4.14}$$

88 Chapter Four

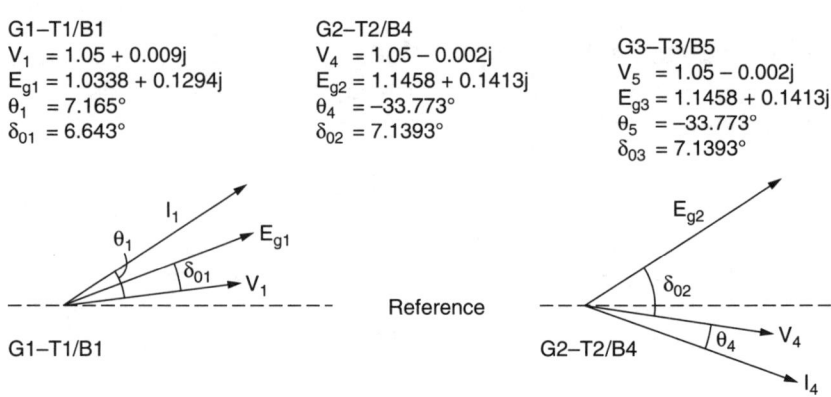

G1–T1/B1
$V_1 = 1.05 + 0.009j$
$E_{g1} = 1.0338 + 0.1294j$
$\theta_1 = 7.165°$
$\delta_{01} = 6.643°$

G2–T2/B4
$V_4 = 1.05 - 0.002j$
$E_{g2} = 1.1458 + 0.1413j$
$\theta_4 = -33.773°$
$\delta_{02} = 7.1393°$

G3–T3/B5
$V_5 = 1.05 - 0.002j$
$E_{g3} = 1.1458 + 0.1413j$
$\theta_5 = -33.773°$
$\delta_{03} = 7.1393°$

FIGURE 4-5 Not-to-scale phasor diagram for G1-T1 and G2-T2.

The conjugate of the power entering the system at bus B4 is:

$$\overline{S}_4 = 1 - 0.666j$$

$V_4 := 1.05 - 0.002j \quad \arg(V_4) = -0.1091° \quad |V_4| = 1.05 \quad \text{referenced to neutral}$

$I_4 = \dfrac{\overline{S}_4}{\overline{V}_4} \quad I_4 := \dfrac{1 - 0.666j}{\overline{V}_4} = 0.951 - 0.636j \quad |I_4| = 1.144 \quad \arg(I_4) = -33.773°$

I_4 lags bus voltage V_4 by 33.773° $\theta_4 := -33.773°$

The emf generated by G2 during steady-state operation is:

$$E_{g2} = V_4 + (X'_{d2})j \times I_4$$

$X'_{d2} = 0.0007 + 0.15$

$E_{g2} = 1.05 - 0.002j + (0.0007 + 0.15j) \times (0.951 - 0.636j) \rightarrow E_{g2} = 1.1458 + 0.1413j$

$E_{g2} := 1.1458 + 0.1413j \quad |E_{g2}| = 1.154 \quad \arg(E_{g2}) = 7.03° \quad \text{referenced to neutral}$

The G2-T2 initial power angle is $\arg(E_{g2}) - \arg(V_4) = 7.13935°$

G2 Initial Power Angle

$$\delta_{02} := 7.1393°$$

The emf generated by G2 leads bus B4 voltage V_4 by 7.1393°. See the phasor diagram in Fig. 4-5.

Transient Stability Problem in a Multimachine Network

G3-T3:
The computations for the G3 generator are identical to the ones for G2. The results are provided below.

$V_5 := 1.05 - 0.002j$ $\arg(V_5) = -0.109°$ $|V_5| = 1.05$ referenced to neutral

$I_5 := \dfrac{1 - 0.666j}{\overline{V_5}} = 0.951 - 0.636j$ $|I_5| = 1.144$ $\arg(I_5) = -33.773°$

I_5 lags bus voltage V_5 by 33.773° $\theta_5 := -33.773°$

$E_{g3} := 1.1458 + 0.1413j$ $|E_{g3}| = 1.154$ $\arg(E_{g3}) = 7.03°$ referenced to neutral

The G3-T3 initial power angle is $\arg(E_{g3}) - \arg(V_5) = 7.13935°$

G3 Initial Power Angle

$$\delta_{03} := 7.1393°$$

The emf generated by G3 leads bus B5 voltage V5 by 7.1393°. See the phasor diagram in Fig. 4-5.

4.5 Network Configuration during the Fault at F1

Consider the three-phase symmetrical short circuit to ground indicated as F1 in Fig. 4.2. During the fault the network looks as illustrated in Fig. 4-6. Table 4-5 provides the per-unit line admittances between busses. For convenience we neglected the resistivity part of the line impedances and the conductance of the admittances and used only the transmission line reactance. In this way we can use the following version of the swing equation:

$$\dfrac{2 \cdot H}{\omega_S} \cdot \dfrac{d^2}{dt^2}\delta = P_S - L - P_e \quad \text{where} \quad P_e = \dfrac{|E_g| \cdot |E_b|}{|x|} \cdot \sin\delta$$

instead of

$$P_e = \dfrac{|E_g| \cdot |E_b|}{|z|} \cdot \cos(\beta - \delta) - \dfrac{|A| \cdot (E_b)^2}{|Z|} \cdot \cos(\beta - \alpha) \qquad \beta \neq \dfrac{\pi}{2}$$

The line admittances neglecting conductances are

$$Y_{1,2} := -3.9j \qquad Y_{1,3} := -3.9j$$
$$Y_{2,4} := -1.997j \qquad Y_{3,5} := -1.997j$$
$$Y_{4,6} := -4.275j \qquad Y_{5,6} := -4.275j$$

The line reactances neglecting resistances are

$$X_{1,2} := 0.256j \qquad X_{1,3} := 0.256j$$
$$X_{2,4} := 0.501j \qquad X_{3,5} := 0.501j$$
$$X_{4,6} := 0.234j \qquad X_{5,6} := 0.234j$$

Figure 4-6 shows the network illustrated in Fig. 4-2 during a three-phase symmetrical fault at F1 with resistances omitted and with generators E_{g2} and E_{g3} lumped together as E_{g23}. Also, it shows several impedance diagrams of the network at different stages of transformation. During a three-phase symmetrical short circuit, the generator output voltages and the line short circuit, currents remain balanced. In fact, the short circuit currents form a balanced system of three phasors shifted 120° from one another, and they cancel at the fault location. Therefore, the potential at F1 is the same as the potential of the neutral, and they can be joined as shown in Fig. 4-6.

Delta-wye conversion and addition of series reactances:

$$0.256j + 0.757j + 0.969j = 1.982j$$

$$0.151j + \frac{0.256j \cdot 0.757j}{1.982j} = 0.249j \qquad 0.07535j + \frac{0.757j \cdot 0.969j}{1.982j} = 0.445j$$

$$\frac{0.969 \cdot 0.256j}{1.982j} = 0.125j$$

Delta-wye conversion:

$$0.249j \cdot 0.445j + 0.445j \cdot 0.125j + 0.125j \cdot 0.249j = -0.198$$

$$\frac{-0.198}{0.125j} = 1.584j \qquad \frac{-0.198}{0.249j} = 0.795j \qquad \frac{-0.198}{0.445j} = 0.445j$$

In the final network configuration of Fig. 4-6, the shunt branches directly across the generators are pure reactance, and as such they

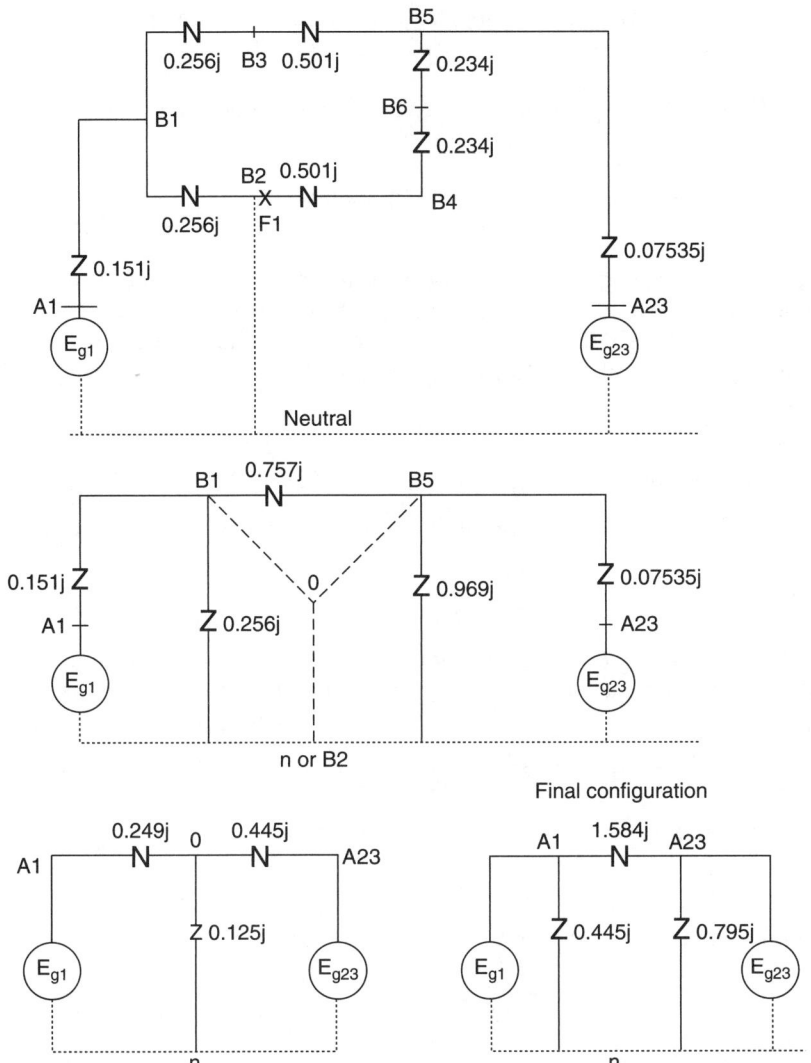

Figure 4-6 Reactance diagrams of the new network configuration during a three-phase fault at F1 and different stages of transformation.

cannot absorb real power. So the only reactance that counts in the evaluation of the flow of real power during transient condition is 1.584j. Besides, it is necessary to verify that generators 2 and 3 are coherent, and therefore they can be lumped together. From Fig. 4-2 and Sec. 4-3, the pretransient, per unit, real electric power delivered by generator G1 is $P_{el} = 1.05$.

From Fig. 4-2 and Section 4-3, the pretransient, per unit, real electrical power delivered by the lumped generator G2/G3 is:

$$P_{e4} + P_{e5} = 2 \quad \text{generator 2 feeds bus 4, and generator 3 feeds bus 5} \quad (4.15)$$

$$P_{e23} = 2 \quad \text{power delivered by lumped generator G2/G3} \quad (4.16)$$

Before the transient, each generator operates in steady state mode at constant power angle and no shaft acceleration. Therefore, the power delivered by the shaft must be equal to the output of electrical power plus the mechanical and electrical losses. Symbolically:

$$P_S = L + P_e$$

$$P_{S1} := 0.02 + 1.05 = 1.07 \quad \text{per unit} \quad 0.02 \cdot 500 = 10 \quad (4.17)$$

$$P_{S23} := 0.02 + 2 = 2.02 \quad \text{per unit} \quad 0.02 \cdot 600 = 12 \quad (4.18)$$

Coherency formula: $\quad \dfrac{P_{sy2}}{M_2} = \dfrac{P_{sy3}}{M_3}$

The synchronizing power coefficient P_{sy} was defined in Eq. (2.45) and M is the inertial constant or the angular momentum at synchronous speed.

$$P_{sy2} = P_{e,max2} \cdot \cos\delta_{02} \qquad P_{sy3} = P_{e,max3} \cdot \cos\delta_{03}$$

From Eq. (2.42) and Fig. 4-2 it is obvious that:

$$P_{e,max2} = P_{e,max3}$$

So the coherency formula in this case is reduced to:

$$\frac{\cos\delta_{02}}{M_2} = \frac{\cos\delta_{03}}{M_3}$$

The generators' initial power angles were computed in Sec. 4.4, and they are identical.

$$\delta_{02} = \delta_{03} = 7.1393° \qquad \cos(7.1393°) = 0.992$$

From Eq. (2.25) we have:

$$M = \frac{G \cdot H}{180 \cdot f} = \frac{300 \cdot 4}{180 \cdot 60} = 0.111$$

Both cosines are identical, and M is the same for both generators, so they are *coherent* and will swing together after a fault or any other disturbance.

Transient Stability Problem in a Multimachine Network

The rating of the lumped generator in 300-MVA base is $G_{23} = 2$ per unit, $H = 8$, and $X_d' = 0.00035$. The lumped transformer rating in 300-MVA base is

$$T_{23} = 2 \text{ per unit}, \quad X_T = 0.075$$

The stored rotational energy at synchronous speed for generator G2 or G3 is

$$GH = 300 \times 4 = 1200 \text{ megawatt-seconds}$$

The stored rotational energy at synchronous speed for the two lumped generators is

$$G_{23} \cdot H_{23} = 1200 \cdot 2 = 2400$$

$$H_{23} = \frac{2400}{600} = 4 \quad \text{600-MVA base}$$

$$H_{23} = 4 \cdot \frac{600}{300} = 8 \quad \text{300-MVA base}$$

The swing equation for the case illustrated in Figs. 4-2 and 4-6 is provided by Eq. (2.36).

$$\frac{2 \cdot H}{\omega_s} \cdot \frac{d^2}{dt^2}\delta = P_a = P_S - L - P_e \quad P_S, L, \text{ and } P_e \text{ in per unit}$$

$$\frac{H}{60 \cdot \pi} \quad H := 8 \quad \frac{8}{60 \cdot \pi} = 0.042 \text{ rad/sec}^2 \quad 0.042 \cdot \frac{d^2}{dt^2}\delta = P_S - L - P_e$$

$$\delta_0 := 7.1393° \quad \text{initial power angle of G2/G3} \quad L := 0.02$$

During steady-state operation or pretransient conditions, the acceleration of the generator's rotor must be zero. Therefore, under these conditions, the mechanical power delivered by the shaft minus the mechanical and electrical losses is equal to the electric output power of the generator. Furthermore, after the fault is cleared, the shaft mechanical power is assumed to be equal to the pretransient value because the fault is cleared in less than 50 millisecond, short enough time intervals for the shaft's rpm to remain almost constant.

$$P_a = P_S - L - P_e = 0 \quad P_S = L + P_e \quad \text{shaft power in steady-state condition, before fault}$$

From Eqs. (4.17) and (4.18) we obtain the per-unit shaft power:

$$P_{S1} = 1.07 \quad P_{S23} = 2.02$$

94 Chapter Four

Equation (2.37) provides the swing equation with the right format for this case.

$$\frac{H}{\pi \cdot f} \cdot \frac{d^2}{dt^2}\delta = P_a = P_s - L - P_e$$

where

$$P_e = \frac{|E_g| \cdot |E_b|}{|X|} \cdot \sin \delta$$

Assuming that the emf of the generator remains constant at the pre-transient value, we get the following value for the per-unit electric real power provided by generator G1 during a fault condition.

$|E_{g1}| = 1.042$ from Sec. 4.4

$$P_{e1d} = \frac{|1.042| \cdot |1.05|}{|1.584j|} \cdot \sin \delta \qquad \frac{|1.042| \cdot |1.05|}{|1.584j|} = 0.691$$

$P_{e1d.max} := 0.691$

$P_{e1d} := 0.691 \cdot \sin \delta$ G1 electric power flow during fault conditions

(4.19)

The per-unit real electric power provided by the lumped generator during fault conditions is

$$P_{e23d} = P_{e2d} + P_{e3d}$$

If we assume that

$P_{e2d} = P_{e3d}$ then $P_{e23d} = 2 \cdot P_{e3d}$

$|E_{g3}| = 1.154$ from Sec. 4.4

$$P_{e3d} = \frac{|1.154| \cdot |1.05|}{|1.584j|} \cdot \sin \delta \qquad \frac{|1.154| \cdot |1.05|}{|1.584j|} = 0.765$$

$P_{e3d,max} = 0.765$ $P_{e3d} = 0.765 \cdot \sin \delta$

$P_{e23d} = 2 \cdot 0.765 \cdot \sin \delta = 1.53 \sin \delta$ $P_{e23d,max} := 1.53$

$P_{e23d} := 1.53 \cdot \sin \delta$ G2/G3 electric power flow during
                                     fault conditions   (4.20)

After the fault at F1 is cleared by the circuit breakers at both ends, (see Figs. 4-6 and 4-7), the per-unit reactance between generators E_{g1} and E_{g23} is

$0.151j + 0.256j + 0.501j + 0.07535j = 0.983j$ $0.256j + 0.501j = 0.757j$

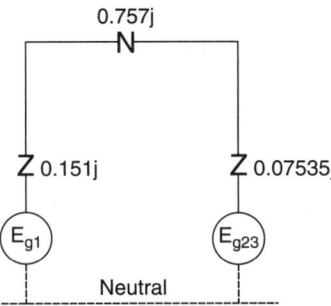

FIGURE 4-7 Reactance between generators after the fault is cleared.

Assuming that the emf of the generators and bus voltages remain constant at the pretransient values, we get the following values for the real electric power provided by the generators after the fault is cleared.

$$P_{e3a,max} := \frac{|1.154| \cdot |1.05|}{|0.983j|} = 1.233$$

$$P_{e23a,max} := 2 \cdot 1.233 = 2.466 \qquad P_{e23a} := 2.466 \cdot \sin \delta \qquad (4.21)$$

$$P_{e1a,max} := \frac{|1.042| \cdot |1.05|}{|0.983j|} = 1.113 \qquad P_{e1a} := 1.113 \cdot \sin \delta \qquad (4.22)$$

Now we can apply the swing equation

$$\frac{2 \cdot H}{\omega_S} \cdot \frac{d^2}{dt^2} \delta = P_a$$

G1 Swing Equation during Fault Conditions
Using Eqs. (4.17) and (4.19), and Table 4-2, we have

$$\frac{10}{60 \cdot \pi} = 0.0531 \qquad P_{S1} = 1.07 \qquad 0.0531 \cdot \frac{d^2}{dt^2} \delta = 1.07 - 0.691 \cdot \sin \delta$$

$$\frac{d^2}{dt^2} \delta = \frac{1}{0.0531}(1.07 - 0.691 \cdot \sin \delta) \qquad \frac{d^2}{dt^2} \delta = 20.15 - 13.013 \cdot \sin \delta$$

$$(4.23)$$

G2/G3 Swing Equation during Fault Conditions
Using Eqs. (4.18) and (4.20) and Table 4-2,

$$\frac{8}{60 \cdot \pi} = 0.0424 \qquad 0.0424 \frac{d^2}{dt^2} \delta = 2.02 - 1.53 \cdot \sin \delta$$

$$\frac{d^2}{dt^2} \delta = \frac{1}{0.0424}(2.02 - 1.53 \cdot \sin \delta) \qquad \frac{d^2}{dt^2} \delta = 47.64 - 36.1 \cdot \sin \delta$$

$$(4.24)$$

G1 Swing Equation after Fault Is Cleared
Using Eqs. (4.17) and (4.22) and Table 4-2,

$$0.0531 \cdot \frac{d^2}{dt^2} \delta = 1.07 - 1.113 \cdot \sin \delta$$

$$\frac{d^2}{dt^2} \delta = \frac{1}{0.0531}(1.07 - 1.113 \cdot \sin \delta) \qquad \frac{d^2}{dt^2} \delta = 20.15 - 20.96 \cdot \sin \delta$$

$$20.15 = 20.96 \cdot \sin \delta \quad \delta = a \sin \frac{20.15}{20.96} = a \sin 0.9613 \quad a \sin 0.9613 = 74.008°$$

(4.25)

G2/G3 Swing Equation after Fault Is Cleared
Using Eqs. (4.18) and (4.21) and Table 4-2,

$$0.0424 \cdot \frac{d^2}{dt^2} \delta = 2.02 - 2.466 \cdot \sin \delta$$

$$\frac{d^2}{dt^2} \delta = \frac{1}{0.0424}(2.02 - 2.466 \cdot \sin \delta) \qquad \frac{d^2}{dt^2} \delta = 47.64 - 58.16 \cdot \sin \delta$$

$$47.64 = 58.16 \cdot \sin \delta \quad \delta = a \sin \frac{47.64}{58.16} = a \sin 0.8191 \quad a \sin 0.8191 = 55°$$

(4.26)

Equations (4.25) and (4.26) indicate that the system could remain stable after the fault is cleared because the accelerating power becomes negative when the power angle of G1 reaches 74° or 55° for G2/G3. We will verify this by solving the swing equation numerically.

4.6 Numerical Solution of the Swing Equation

$$\frac{H}{60 \cdot \pi} \frac{d^2}{dt^2} \delta = P_s - L - P_e \qquad \text{swing equation} \qquad L := 0.02$$

Losses are assumed constant. These losses have a damping effect in the transient condition, because it consumes the energy of the transient.

Swing Equation for G1 during Fault Conditions
From Eq. (4.23) we get:

$$\frac{d^2}{dt^2} \delta = 20.15 - 13 \cdot \sin \delta$$

Transient Stability Problem in a Multimachine Network

The swing equation can be written as:

$$\frac{d}{dt}\left(\frac{d}{dt}\delta\right) = 20.15 - 13 \cdot \sin\delta$$

Define two new functions $\delta_0(t)$ and $\delta_1(t)$:

$$\delta_0(t) = \delta(t) \quad \text{and} \quad \delta_1(t) = \frac{d}{dt}\delta_0(t)$$

Now the original differential equation can be written as:

$$\frac{d}{dt}\delta_1(t) = 20.15 - 13 \cdot \sin\delta_0(t)$$

This new differential equation has two functions, $\delta_0(t)$ and $\delta_1(t)$, instead of one, $\delta(t)$. These two new functions are related by the following equation:

$$\delta_1(t) = \frac{d}{dt}\delta_0(t)$$

The original differential equation has been converted to a system of two differential equations, written with the derivatives alone on the left-hand side of the equal sign:

$$\frac{d}{dt}\delta_0(t) = \delta_1(t)$$

$$\frac{d}{dt}\delta_1(t) = 20.15 - 13 \cdot \sin\delta_0(t)$$

Swing Equation for G2/G3 during Fault Conditions

From Eq. (4.24) we get

$$\frac{d^2}{dt^2}\delta = 47.6 - 36.1 \cdot \sin\delta$$

The swing equation can be written as

$$\frac{d}{dt}\left(\frac{d}{dt}\delta\right) = 47.6 - 36.1 \cdot \sin\delta$$

Define two new functions, $\delta_2(t)$ and $\delta_3(t)$, such that

$$\delta_2(t) = \delta(t) \quad \text{and} \quad \delta_3(t) = \frac{d}{dt}\delta_2(t)$$

Chapter Four

Now the original differential swing equation for G2/G3 is:

$$\frac{d}{dt}\delta_3(t) = 47.6 - 36.1 \cdot \sin \delta_2(t)$$

This new differential equation has two functions, $\delta_2(t)$ and $\delta_3(t)$, instead of one, $\delta(t)$. These two new functions are related by the following equation:

$$\delta_3(t) = \frac{d}{dt}\delta_2(t)$$

The original differential equation has been converted to a system of two differential equations:

$$\frac{d}{dt}\delta_2(t) = \delta_3(t) \qquad \frac{d}{dt}\delta_3 = 47.6 - 36.1 \cdot \sin \delta_2(t)$$

Define a derivative vector D containing the four new functions of t for the numerical solver. Use the MathCad array subscript to indicate the derivative order.

$$D(t, \delta) := \begin{bmatrix} \delta_1 \\ 20.15 - 13 \cdot \sin[\delta_0(t)] \\ \delta_3 \\ 47.6 - 36.1 \cdot \sin(\delta_2) \end{bmatrix} \qquad D = \begin{pmatrix} \frac{d}{dt}\delta_0 \\ \frac{d}{dt}\delta_1 \\ \frac{d}{dt}\delta_2 \\ \frac{d}{d}\delta_3 \end{pmatrix}$$

Let us use the Rkadapt solver to solve the system of second order differential equations:

$t_1 := 0 \qquad \delta_0 := 0.116 \qquad \delta_1 := 0 \qquad 20.15 - 13 \cdot \sin(\delta_0) = 18.65$

$\qquad\qquad \delta_2 := 0.125 \qquad \delta_3 := 0 \qquad 47.6 - 36.1 \cdot \sin(\delta_2) = 43.1$

$ic := \begin{pmatrix} 0 \\ 18.65 \\ 0 \\ 43.1 \end{pmatrix} \qquad t1 := 0 \qquad t2 := 1 \qquad npoints := 1000$

$Q := Rkadapt(ic, t1, t2, npoints, D)$

Transient Stability Problem in a Multimachine Network

$$Q = \begin{array}{|c|c|c|c|c|c|} \hline & t & \delta_1 & \frac{d}{dt}\delta_1 & \delta_3 & \frac{d}{dt}\delta_3 \\ \hline & 0 & 1 & 2 & 3 & 4 \\ \hline 50 & 0.05 & 0.958 & 19.647 & 2.189 & 44.181 \\ \hline 51 & 0.051 & 0.977 & 19.667 & 2.233 & 44.2 \\ \hline 52 & 0.052 & 0.997 & 19.686 & 2.277 & 44.219 \\ \hline 53 & 0.053 & 1.017 & 19.706 & 2.321 & 44.24 \\ \hline 54 & 0.054 & 1.036 & 19.725 & 2.366 & 44.62 \\ \hline 55 & 0.055 & 1.056 & 19.745 & 2.41 & 44.284 \\ \hline 56 & 0.056 & 1.076 & 19.764 & 2.454 & 44.309 \\ \hline 57 & 0.057 & 1.096 & 19.783 & 2.499 & 44.334 \\ \hline 58 & 0.058 & 1.115 & 19.803 & 2.543 & 44.36 \\ \hline 59 & 0.059 & 1.135 & 19.822 & 2.58 & 44.388 \\ \hline 60 & 0.06 & 1.155 & 19.841 & 2.632 & 44.418 \\ \hline 61 & 0.061 & 1.175 & 19.86 & 2.676 & 44.448 \\ \hline 62 & 0.062 & 1.195 & 19.88 & 2.721 & 44.48 \\ \hline 63 & 0.06 & 1.215 & 19.899 & 2.765 & 44.514 \\ \hline 64 & 0.064 & 1.235 & 19.918 & 2.81 & 44.549 \\ \hline 65 & 0.065 & 1.254 & 19.937 & 2.854 & ... \\ \hline \end{array}$$

<<<< 0.05, 0.958, 19.647, 2.189, 44.181

$G1 := 1.67 \quad H1 := 10 \quad f := 60$

$M_1 = \dfrac{G1 \cdot H1}{\pi \cdot f} = 0.089$

$G23 := 2 \quad H23 := 8$

$M_{23} := \dfrac{G23 \cdot H23}{\pi \cdot f} = 0.085$

$M_1 \cdot Q^{(2)}$ power angle acceleration

[1]Only 16 points are shown in Table 4-7, but in MathCad you could scroll down to any of 1000 points.

TABLE 4-7 Matrix of Power Angles and Their Acceleration during Fault Conditions[1]

We do not consider the system stability during fault conditions because the fault will be cleared in 50 milliseconds. However, Table 4-7 at 50 ms provides the initial conditions to numerical solve the differential equations after the fault has been cleared. This will tell us if the generators would remain stable after clearing the fault.

At 50 milliseconds: $\delta_1 = 0.958 \text{ rad} \quad \dfrac{d}{dt}\delta_1 = 19.65 \quad \delta_3 = 2.189$

$\dfrac{d}{dt}\delta_3 = 44.181$

$\text{ica} := \begin{pmatrix} 0.958 \\ 19.65 \\ 2.189 \\ 44.18 \end{pmatrix}$

$Q_a := \text{Rkadapt}(\text{ica}, t1, t2, \text{npoints}, D)$

100 Chapter Four

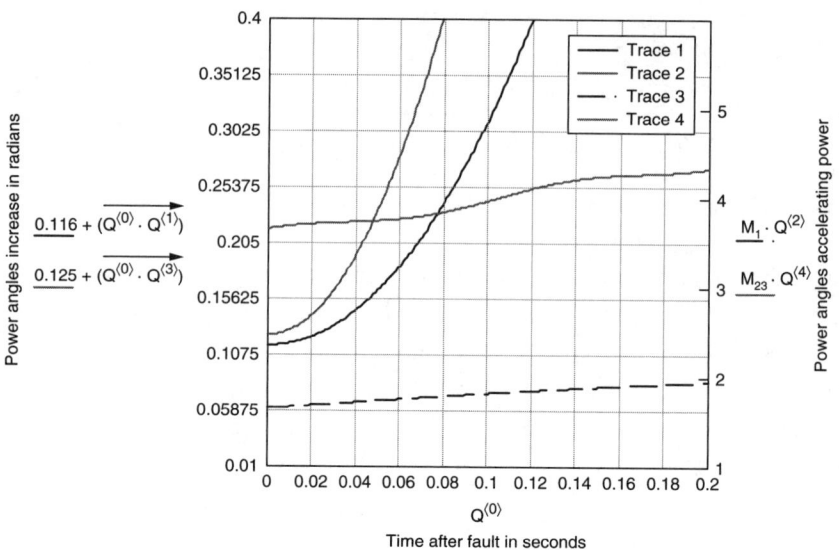

FIGURE 4-8 Power angles in radians and their accelerating power in per unit during fault.

Figure 4-9 shows that the three generators will become unstable after clearing the fault because the power angles will increase without limit pushed by an increasing or almost constant accelerating power.

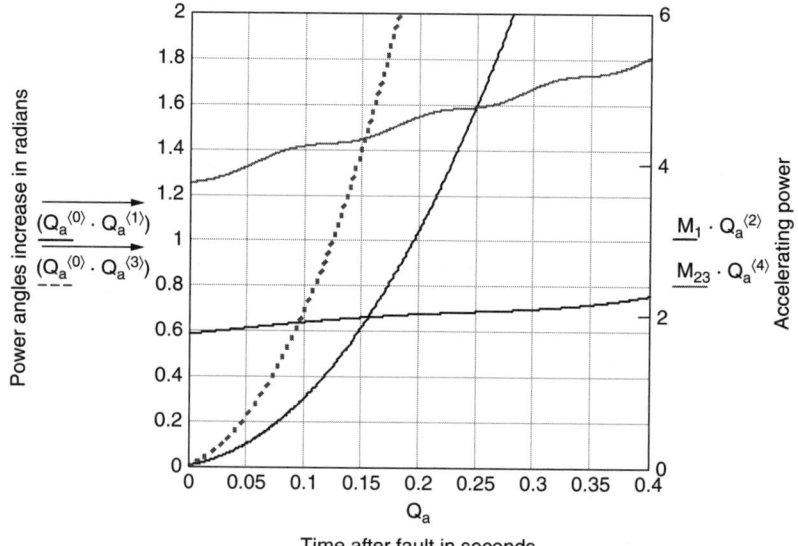

FIGURE 4-9 Power angles and their accelerating power in per unit after the fault has been cleared.

Transient Stability Problem in a Multimachine Network

$$Q_a = \begin{array}{|c|c|c|c|c|c|} \hline & t & \delta_1 & \frac{d}{dt}\delta_1 & \delta_3 & \frac{d}{dt}\delta_3 \\ \hline & 0 & 1 & 2 & 3 & 4 \\ \hline 0 & 0 & 0.958 & 19.65 & 2.189 & 44.18 \\ \hline 1 & 1 \cdot 10^{-3} & 0.978 & 19.67 & 2.233 & 44.199 \\ \hline 2 & 2 \cdot 10^{-3} & 0.997 & 19.69 & 2.277 & 44.218 \\ \hline 3 & 3 \cdot 10^{-3} & 1.017 & 19.71 & 2.322 & 44.239 \\ \hline 4 & 4 \cdot 10^{-3} & 1.037 & 19.73 & 2.366 & 44.261 \\ \hline 5 & 5 \cdot 10^{-3} & 1.057 & 19.751 & 2.41 & 44.284 \\ \hline 6 & 6 \cdot 10^{-3} & 1.076 & 19.771 & 2.454 & 44.308 \\ \hline 7 & 7 \cdot 10^{-3} & 1.096 & 19.791 & 2.499 & 44.333 \\ \hline 8 & 8 \cdot 10^{-3} & 1.116 & 19.811 & 2.543 & 44.36 \\ \hline 9 & 9 \cdot 10^{-3} & 1.136 & 19.831 & 2.587 & 44.388 \\ \hline 10 & 0.01 & 1.156 & 19.851 & 2.632 & 44.417 \\ \hline 11 & 0.011 & 1.175 & 19.871 & 2.676 & 44.448 \\ \hline 12 & 0.012 & 1.195 & 19.891 & 2.721 & 44.48 \\ \hline 13 & 0.013 & 1.215 & 19.911 & 2.765 & 44.513 \\ \hline 14 & 0.014 & 1.235 & 19.931 & 2.81 & 44.548 \\ \hline 15 & 0.015 & 1.225 & 19.951 & 2.854 & \ldots \\ \hline \end{array}$$

TABLE 4-8 Matrix of Power Angles and Their Acceleration after Clearing the Fault

G1 Rotor Natural Frequency and Period of Oscillation

Equation (2.52) provides the undamped natural frequency of oscillation of each generator rotor, which for convenience is repeated here:

$$\omega_n = \sqrt{\frac{\omega_s \cdot P_{sy}}{2 \cdot H}} \qquad f_n = \frac{1}{2 \cdot \pi}\sqrt{\frac{\omega_s \cdot P_{sy}}{2 \cdot H}} \qquad T_n = \frac{1}{f_n}$$

From Eq. (2.45) we obtain the synchronizing power coefficient.

$$P_{sy} = P_{e,max} \cdot \cos\delta \qquad \delta = \text{operating power angle just after fault is cleared} \qquad (4.27)$$

$\delta_1 := 0.958$ rad   See initial conditions at 50 milliseconds in Table 4-7 or Table 4-8 at t = 0.

$P_{sy1} := 1.113 \cdot \cos\delta_1$ See Eqs. (4.22) and (4.27) $P_{sy1} = 0.64$

$$\omega_{nl} := \sqrt{\frac{377 \cdot 0.64}{2 \cdot 10}} = 3.47 \text{ rad/sec} \qquad \omega_{nl} = 2 \cdot \pi \cdot f_n$$

$$f_{nl} = \left(\frac{3.47}{2 \cdot \pi}\right) = 0.552 \text{ Hz} \qquad T_{nl} := \frac{1}{0.552} = 1.812 \text{ seconds}$$

Any one of the three generators could experience transient-induced low-frequency oscillations, and if the natural frequency of oscillation of any of the generator's rotors is in the same range as the transient-induced oscillations, then the rotor could be severely damaged. All generator rotors should be dynamically balanced to avoid natural frequency of oscillations close to the expected transient-induced oscillations.

The vast network of interconnected electric power systems in the United States and Canada contains hundreds of AC generators. The study of the transient stability of this enormous power system is a giant task that requires the collaboration of all the interconnected utilities. I suggest the following:

1. Divide the system into sections. In steady-state condition, each section must be quasi-independent of the others; that is, it should never import more than 15 percent of the section maximum demand.

2. In all the tie lines with other sections, the synchronizing power coefficient must be positive in both directions and close in value, with no more than 30 percent difference.

3. Coherent generators within a section must be lumped together for analysis.

4. Numerically, solve the swing equation for each section, for the first swing or 1 min. The D vector should have twice the number of rows as generators in the section.

5. Once it is determined that by themselves all the sections are transient stable, then the stability of the entire system could be considered.

CHAPTER 5
High-Voltage AC Capacitors

During charging and discharging events, capacitor banks could introduce very large oscillating voltages and currents of nonharmonic frequencies into the power system. This chapter covers the dynamic of these events for high-voltage capacitor banks. We analyzed the discharge current event using the classical solution of the second-order linear differential equation representing a capacitor discharge event in which the applied AC voltage was suddenly removed from an *RLC* series circuit due to a single line-to-ground fault. The nonharmonic, 351-hertz (Hz) discharge current starts at very high amplitude, 4561 amperes (A), and 0.8 second later is down to 617 A. The second-order differential equation representing a charging event in the same circuit, after the successful auto reclosing of the line breaker, was solved using a numerical method of analysis. The charging current rises very quickly and reaches 27,000 A in less than 1 second. This current is composed of high-frequency and low-frequency components; the low-frequency component is nonharmonic at 2.27 Hz. The peak value of the voltage oscillation is 13,505 kilovolts (kV), which is 47 times the rated line-to-neutral voltage of the power system. The charging event is practically finished after four seconds.

5.1 Introduction

A capacitor essentially consists of two conductive surfaces placed a short distance apart and kept separate by a thin sheet of insulation material called *dielectric*, which occupies the space between the conductive surfaces. The properties of capacitors are a function of their geometric configuration, the dielectric used, the material of the conductive surfaces, and the connecting contacts. In high-voltage applications, individual capacitors may be electrically connected in series or parallel and physically stacked on top of each other. Keep in mind that the thin dielectric needs to withstand the line high voltage. Actually, this is one reason why they are stacked and forced to share the stress of the high voltage evenly. Throughout the manufacturing process, great care and attention are directed to avoid air bubbles, humidity, and impurity's contamination, especially during the fabrication and installation of the dielectric material and final assembly. Ruby mica and Pyrex glass are excellent dielectric materials, extremely reliable, stable, and tolerant to high temperatures and voltages. For small values of capacitance, in the picofarad range, sometimes, depending on the circuit voltage, ceramic disks with metallic coatings are used with wire leads bonded to the coating. Larger capacitance values can be achieved by multiple stacks of disks.

When a capacitor is charged, there is not any charge of electricity added to the plates. What happens is that the space relations between protons and electrons become rearranged (charge separation) so that an electrical stress, called an *electrical field*, is created between the plates. In fact, when an electric circuit is connected to the opposite plates of the capacitor, it will give back the energy stored in the electrical field minus the internal dielectric losses and the energy loss in the circuit resistance.

According to their use, high-voltage AC capacitors are classified as follows:

- *Transient recovery voltage capacitors (TRVs).* These high-voltage capacitors are used in combination with circuit breakers to attenuate the transient overvoltages that may occur during switching or interruption operations.

- *Coupling capacitors (CCs).* They are used to couple signals from high-frequency carriers to transmission lines.

- *Grading capacitors.* These capacitors are used in parallel with the interrupting units, in multibreak circuit breakers to distribute equally the voltage across all contact points during switching operations.

- *Power factor correction capacitors.* These are also called *reactive power compensation capacitors*. This classification is applicable to a wide range of voltage levels from 120-volt (V) motors to 500-kV transmission lines or substations. The purpose of these capacitors is to reduce the demand of reactive power and therefore save megavolt-amperes (MVA) of generator capacity. Usually, these capacitors are rated not in picofarads but rather in reactive volt-amperes.

- *Capacitor voltage transformers.* These capacitors are a cheap substitution for wound-measuring potential transformers. These capacitors are used as a voltage divider; they are series-connected to ground. A low-voltage transformer is usually connected to a tap in the series to provide a secondary voltage that could be used for metering purposes.

5.2 Capacitor Steady-State Equations

The basic equation that describes capacitor steady-state operation is

$$q = C \times v_c \tag{5.1}$$

where C = capacitance, in farads (F)
v_c = voltage across plates, in volts
q = charge on positive plate, coulombs (C)

The charge on the capacitor plates is the result of a current circulating in the circuit connecting the plates. Equation (5.1) expresses that the charge is proportional to the voltage between the plates. And we know that the charging current is equal to the rate of change of the charge on the plates. Differentiating Eq. (5.1), we get Eq. (5.2):

$$i = \frac{d}{dt} q = \frac{d}{dt}(C \times v_c) = C \times \left(\frac{d}{dt} v_c\right) \quad i = C \times \left(\frac{d}{dt} v_c\right) \tag{5.2}$$

Equation (5.2) shows that the current in a capacitor is proportional to the rate of change of the voltage across the capacitor. The energy stored in the electrical field of capacitors is provided by Eq. (5.3).

$$W = \frac{C \times (v_c)^2}{2} \tag{5.3}$$

where W = energy, in joules
C = capacitance, in farads
v_c = voltage across plates, in volts

This energy is continuously interchanged between the capacitor's electrical field and the circuit inductance magnetic field. For this interchange to take place, it is necessary to have a circulating current. If the driving voltage were removed, the interchange would continue until all the energy was consumed in the resistance of the circuit.

5.3 Basic Capacitor Connections

Capacitors Connected in Series

The formula to compute the total capacitance of any number of capacitors connected in series is

$$\frac{1}{C_T} = \frac{1}{C_1} + \frac{1}{C_2} + \frac{1}{C_3} + \cdots$$

$$C_T = \frac{1}{1/C_1 + 1/C_2 + 1/C_3 + \cdots}$$

The formula for just two capacitors connected in series is

$$C_T = \frac{C_1 \times C_2}{C_2 + C_1}$$

Capacitors Connected in Parallel

The formula to compute the total capacitance is

$$C_T = C_1 + C_2 + C_3 + \cdots$$

Capacitance from $KVAR_c$ to Picofarads

High-voltage power system capacitors are usually rated in reactive kilovolt-amperes (KVAR), the formula to calculate the actual capacitance in picofarads that corresponds to a given KVAR is deduced below. The deduction is per phase, and $KVAR_c$ is the actual operating reactive kilovolt-amperes of the capacitor, which often is different from the rating on the capacitor plate.

$$KVAR_c = kV_c \times I_c = kV_c \times \frac{10^3 \times kV_c}{X_c} = \frac{10^3 \times (kV_c)^2}{X_c} \quad (5.4)$$

$$X_c = \frac{10^3 \times (kV)^2}{KVAR_c} \quad \text{ohms} \quad (5.5)$$

where kV_c = actual kilovolts across capacitor, kV
$KVAR_c$ = actual reactive kilovolt-amperes of the capacitor
X_c = capacitor reactance, in ohms

$$\frac{1}{\omega \times C \times 10^{-12}} = \frac{10^3 \times (kV_c)^2}{KVAR_c}$$

C is in picofarads.

$$C = \frac{10^9 \times KVAR_c}{2\pi f_c \times (kV_c)^2} \quad \text{picofarads (pF)} \quad (5.6)$$

Equation (5.6) provides the formula to calculate capacitance in picofarad from $KVAR_c$ computed Eq. (5.7).

The formula to calculate the actual operating rating from the rating on the plate is

$$KVAR_c = KVAR \times \left(\frac{kV_c}{kV}\right)^2 \times \left(\frac{f_c}{f}\right) \quad (5.7)$$

where KVAR = rated KVAR given on capacitor plate
kV = rated kilovolts on capacitor plate
kV_c = actual kilovolts across capacitor
f = frequency given on capacitor plate
f_c = operating frequency

5.4 Reactive Power Compensation

In general, most of the loads connected to an electric power system are inductive loads, and the current supply to these inductive loads lags the line voltage. The reactive power supplied to an inductive load is, by convention, considered as positive, and the reactive power *supplied* to a capacitive circuit is considered negative. However, for utility engineers, capacitors are a means of *delivering* KVAR (reactive kilovolt-amperes) at the point of installation, and they interpret this as positive delivered reactive power. So, it is reasonable to think that capacitors provide the reactive power required by the inductive load. What happens is that the reactive power interchange between the magnetic field of the load inductance and the electrical field of the capacitor occurs closer to the load. Without capacitors the load reactive power (magnetizing current) would need to be provided by the generator. This is called *reactive power compensation*. Figure 5-1 is an illustration of the power triangle showing the components that makeup the complex power or total volt-amperes. Reactive power compensation reduces the total current output of the generator and in this way makes more amperes available to provide real power. It also decreases the current flow in transmission and distribution lines, in transformers, and in interruption devices, such as breakers, and therefore reduces the copper loss in all of them. Any device connected before the point of capacitor installation benefits from these reductions. Utilities promote the installation of power factor correction capacitors at the consumer site by penalizing them if the power factor falls below a set value. In summary, the reactive power compensation releases a great amount of the system capacity, which was dedicated to supply the reactive part of the load, reduces the losses, and improves the voltage profile. However, capacitor compensation could introduce new problems—some of them hard to track and analyze. Many of them occur during charging capacitor banks, especially after breaker reclosing operations. During the charging (it could occur also during discharging) transient period, capacitors could introduce voltage oscillations of very large amplitude, much larger than the line-rated

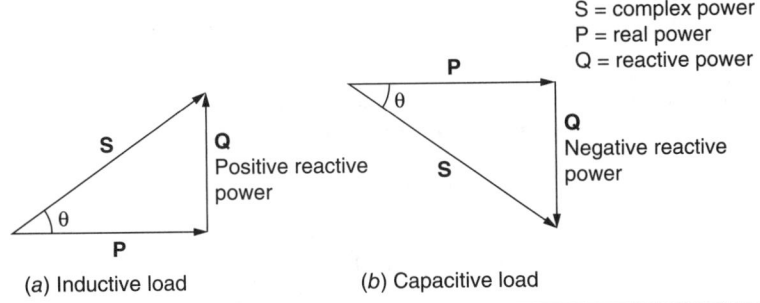

FIGURE 5-1   Power triangle.

voltage; these voltage oscillations could punch through the insulator material of some devices or flash over the line insulators. These oscillations also produce vibration in motors and generators, and if the frequencies of these voltage oscillations are subharmonic (smaller than 60 Hz) and equal or close to one of the torsion frequencies of the shaft that mechanically couples the driving turbine with the AC generator, they could damage the machine. Reactive power compensation in electric power systems is accomplished by either series-or shunt-connected capacitor banks. The ideal compensator would have only capacitors, no transformers, no reactors, and no switching thyristors. However, in high-voltage applications, transformers are frequently used to reduce the line voltage. Capacitors rated at lower voltage are cheaper, less bulking, and easier to install. To reduce the possibility or magnitude of the voltage oscillations, avoid the use of reactors and transformers, and anything else having inductance, in capacitor installations. At relatively low-voltage applications, some reactive power compensators use thyristors as switching devices. The reader should keep in mind that their fast switching could introduce high-frequency oscillations in the power system.

5.5 Series-Connected Capacitor Banks

Reactive power compensation with series-connected capacitor banks is primarily installed in substations to compensate high-voltage transmission lines. A series capacitor bank releases a significant amount of the capacity of the transmission line and generators that were previously dedicated to supply the reactive part of the load. It also reduces the losses and improves the voltage profile. Unfortunately, during discharging and charging events, series-connected capacitors are prone to introduce into the power system low-frequency voltage oscillations that could produce what is called *subsynchronous resonance* (SSR) in some of the connected synchronous rotary equipment. For the case of series compensated radial transmission systems, the undamped natural frequency of oscillation introduced by the capacitor-inductance combination is

$$f_n = f_s \times \sqrt{\frac{X_C}{X_L}} \quad \text{natural frequency of oscillation}$$

where the electrical system synchronous frequency is within the range of $57 \leq f_s \leq 63$. The complement frequency is defined as

$$f_c = f_s - f_n$$

Complement frequency $\quad f_c = f_s \left(1 - \sqrt{\frac{X_C}{X_L}}\right) \quad$ (5.8)

Subsynchronous resonance occurs when the complement frequency is equal or close to one of the torsion frequencies of the turbine-generator system. In this case, the low-frequency oscillation goes through the stator's windings and induces a small subsynchronous voltage in the rotor field winding and in the shaft itself, which produces large subsynchronous current in the rotor's winding, due to the low reactance at subsynchronous frequency. It also creates circulating currents in the rotor's shaft. These currents introduce oscillatory components in the generator rotor torque whose phase is such (in sync) that it increases the rotor vibrations. If the mechanical energy lost by the shaft as it oscillates is not large enough, the oscillatory component of the torque will grow and could cause severe damage to the shaft and bearings.

Capacitor Bank Connected in Series with the Line

A series capacitor bank consists of many single capacitors (units) that could be series- or parallel-connected until one obtains the required KVAR and the necessary voltage rating which together with the insulated platform rating should withstand the line voltage to ground in normal operating conditions, including the expected current and voltage fluctuations due to sudden load changes. For overvoltage protections during transient conditions produced by line faults or due to capacitor charging or discharging, several protecting devices are used. In the typical installation shown in Fig. 5-2, the protection function is as follows: The bank of metal oxide varistors has very high resistance at normal operating conditions, but once the voltage across the capacitor bank reaches the protected level, its resistance becomes much lower and practically shorts out the capacitor bank. However, if the line voltage oscillations across the capacitor bank (the MOV resistance is not low enough) become

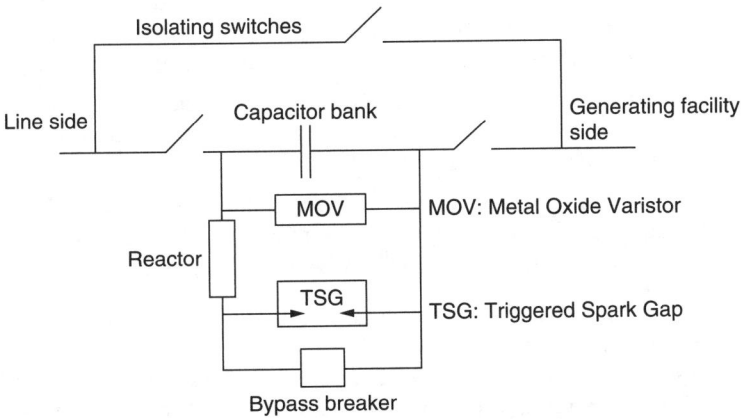

FIGURE 5-2 One-phase illustration of a series capacitor bank.

very high due to the high current produced by a nearby fault, then the combination of reactor and triggered spark gap enters in action and bypasses both the capacitor bank and the metal oxide varistor (MOV). The reactor damps the oscillations produced by the sparks in the TSG device and also limits the capacitor discharge current produced by the shorting out of the bank. If everything else fails, the breaker will close and bypass the capacitor bank for good.

Often the entire capacitor bank installation is mounted on a platform insulated to withstand the line voltage to ground plus any voltage surge. Each phase is isolated or mounted independently of the other (two) phases. The bank should be protected against overvoltages and provided with a set of isolating switches and bypassing breakers. Figure 5-2 provides an illustration of the connections for one phase of a typical three-phase installation.

As shown in Fig. 5-2, the series capacitor bank includes overvoltage protection, which in this specific case consists of a zinc oxide varistor bank connected parallel with the capacitors, a damping reactor, a triggered spark gap, and a bypass breaker.

5.6 Shunt-Connected Capacitor Banks

Shunt capacitor banks are installed everywhere in the power system: at substation busses feeding transmission or distribution lines, along lines, and at the user site. They are an excellent way of supplying KVAR at the point of installation. In addition they could be connected to the high-voltage power lines using step-down transformers (you cannot do that with series-connected capacitor banks). In this way the capacitors and switching equipment (if any) could be rated at the lower voltage in the secondary side of step-down transformers.

The reasons for installing shunt capacitor banks are to

- Reduce the line voltage drop by decreasing the line current and its lagging angle (load KVARs are inductive in general). This raises the voltage at the point of capacitor installation.

- Reduce the power losses in the line and installed equipment before the point of installation of capacitor bank. Less current means less heat dissipated in resistance.

- Improve the effective use of the generating equipment MVA capacity. Capacitors reduce, the reactive power delivered by generators, and increase the available real power.

Figure 5-3 illustrates the connections for one phase of a typical three-phase shunt capacitor installation.

There are many versions of static voltage compensators, and Fig. 5-3 is only an illustration. One important thing that must be verified is that the resultant voltage wave shape does not have a DC component. For this to happen, thyristor 1 must come off at the exact

High-Voltage AC Capacitors

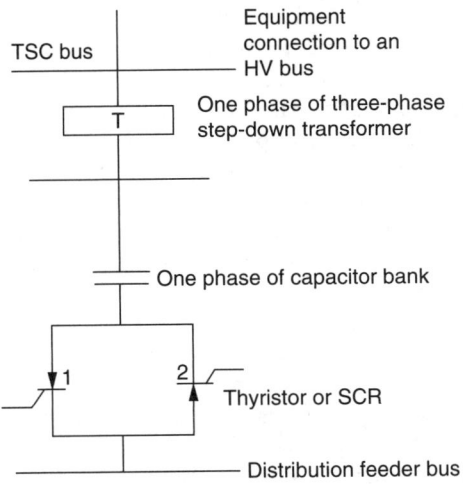

FIGURE 5-3 Thyristor switched shunt-connected capacitor bank.

instant that thyristor 2 comes on, and vice versa. This must happen in all the phases. Harmful switching transients are injected into the system when a thyristor is switched off while it's still conducting current. To avoid transients, the thyristor should be turned off or turned on while the instantaneous voltage across the capacitor is in a positive or negative peak. There are many ways to adjust the reactive power compensation to the changes in load demand. Keep in mind that the load is always inductive and some amount of correction is always required. The adjustment can be done by switching, in or out, capacitor bank steps manually or via a clock-controlled switch. Or better yet, it can be done by electronically and automatically increasing or decreasing the correct amount of compensation. Figure 5-4 shows the basis of one version to accomplish this.

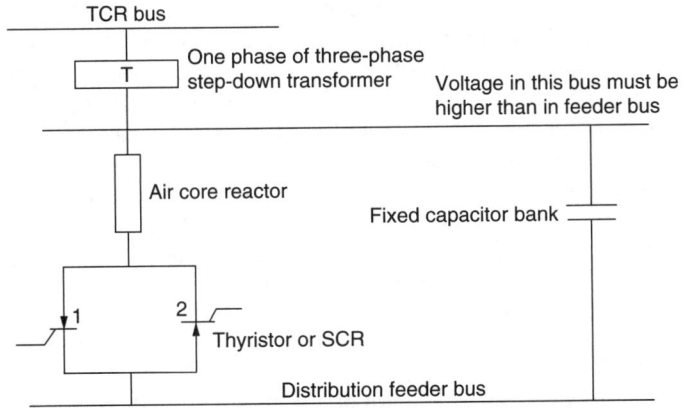

FIGURE 5-4 One phase of a thyristor controlled reactor (TCR).

In Fig. 5-4 the capacitor bank delivers to the distribution feeder a fixed amount of negative capacitive reactive power, and the thyristors control the amount of positive inductive reactive power that the reactor delivers to the distribution bus. The difference is the net amount of reactive power compensation. The inductive KVAR delivered by the reactor branch is controlled by changing the thyristor's current conduction period in each cycle, from zero conduction, with the gate firing signal off when the thyristor is ready to conduct, to full conduction that occurs when there is no intentional delay in the gate's firing signals. The design of the reactor in the system shown in Fig. 5-4 is such that at full thyristor current conduction the inductive KVAR delivered by the inductive branch is equal to the capacitive KVAR provided by the capacitor bank and there is no reactive power compensation. If some compensation is required, then the gate firing signal is delayed; this decreases the inductive KVAR delivered and increases the net capacitive reactive power compensation. The maximum capacitive reactive power compensation occurs when there is no current conduction through the thyristors, with the gate firing signals off.

5.7 AC Voltage Suddenly Applied To or Removed From an RLC Series Circuit

The deenergization and subsequent energization of entire sections of an electric power system, programmed or not, are events sure to occur. Affected capacitor banks will generate transient oscillating voltages in both occurrences. The sudden deenergization will produce nonharmonic oscillations of low or high frequency (lower or higher than 60 Hz) depending of the value of the circuit parameters. The 60-Hz fundamental would not be included in the spectrum because there is no applied voltage. The energization case is more severe, and it contains the 60-Hz fundamental.

Capacitor banks are either series- or shunt-connected to the system. When a short circuit to ground occurs in an electric power system, all directly affected shunt-connected capacitors could suddenly discharge—and then charge again following an automatic reclosing operation. Most times, these events will introduce dangerous low-frequency current and voltage oscillations in the power system. These oscillations could not be harmonics of the 60-Hz fundamental and could produce mechanical vibrations in all motors and generators connected to the power system. If these vibrations are equal or close to the natural frequency of oscillation of a machine shaft, it could even break or damage it.

Example 5-1
The requirement is for installation of a shunt-connected capacitor bank of 300 MVA of reactive power connected at the midpoint of a short

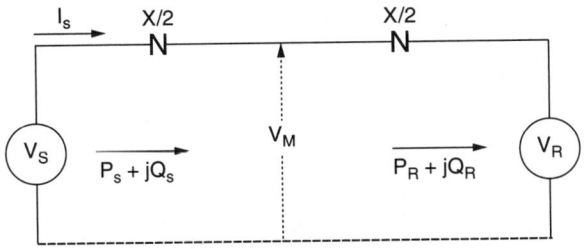

FIGURE 5-5 Lossless short transmission line.

[40 miles (mi) long] 500-kV transmission line. The load at the receiving end is a balanced three-phase load; the line resistance and capacitance to earth are neglected (lossless transmission line). Figure 5-5 is the single-phase equivalent circuit of the short transmission line.

The capacitor bank consists of a neutral grounded three-phase wye with 100 MVAR per phase. It is necessary to check the induced line voltage oscillations and capacitor bank discharging and charging currents after a single line-to-ground fault. The first reclosing operation of the line breaker on the sending end takes place 0.2 second after the first tripping, and the second (if necessary) reclosing takes place 6 seconds after the second tripping.

$$\text{Line reactance} = 1\ \Omega/\text{mile} \qquad X := 40\ \Omega$$

$$\frac{X}{2} = 20\ \Omega \qquad I_S = I_R$$

Let the line-to-neutral voltage at the sending and receiving ends, both referred to the transmission line neutral, be

$$j := \sqrt{-1}$$

$$V_S := \frac{500 \cdot 10^3}{\sqrt{3}}[\cos(0.8727) + \sin(0.8727) \cdot j] \qquad V_R := \frac{498 \cdot 10^3}{\sqrt{3}}[\cos(0.5236) + \sin(0.5236) \cdot j]$$

$$\frac{500 \cdot 10^3}{\sqrt{3}} = 288675 \qquad \frac{498 \cdot 10^3}{\sqrt{3}} = 287520$$

$$V_S = 185549 + 221145j\ \text{volt} \qquad V_R = 249000 + 143761j\ \text{volt}$$

The line current is then

$$I_S := \frac{V_S - V_R}{X \cdot j} = 1935 + 1586j\ \text{amp} \qquad I_R = 1935 + 1586j\ \text{amp}$$

$$|1935 + 1586j| = 2502 \qquad \operatorname{atan}\left(\frac{1586}{1935}\right) = 687 \times 10^{-3}\ \text{rad} \qquad 0.687\ \text{rad} = 39°$$

The instantaneous line current is

$$i_S = \sqrt{2} \cdot 2502 \cdot \sin(\omega \cdot t + 0.687)$$

$$\overline{I_S} = 1935 - 1586j \quad \overline{I_R} = 1935 - 1586j$$

Sending end complex power:

$$P_S + Q_S \cdot j = V_S \cdot \overline{I_S} \quad V_S \cdot \overline{I_S} = 710 \times 10^6 + 133j \times 10^6$$

133 MVA per phase of reactive power sent by generators

Receiving end complex power:

$$P_R - Q_R \cdot j = \overline{V_R} \cdot I_R \quad I_R := I_S \quad \overline{V_R} \cdot I_R = 710 \times 10^6 + 117j \times 10^6$$

117 MVA per phase of reactive power delivered to the load. Real power delivered is equal to the real power sent.

The reactive power absorbed by the line is

$$Q_L = Q_S - Q_R = 133 - 117 = 16 \text{ MVA}$$

The midpoint line-to-neutral voltage is

$$V_M := V_S - I_S \cdot 20 \cdot j = 217 \times 10^3 + 182j \times 10^3 \tag{5.9}$$

$$|V_M| = 284 \times 10^3 \text{ volt} \quad \operatorname{atan}\left(\frac{182}{217}\right) = 0.698 \text{ rad} \quad 0.698 \text{ rad} = 40°$$

The instantaneous midpoint line-to-neutral voltage is

$$V_M = \sqrt{2} \cdot 284 \cdot 10^3 \cdot \sin(\omega \cdot t + 0.698) \tag{5.10}$$

Capacitance from KVAR$_c$ to Picofarads

The load requires 117,000 KVAR per phase; the capacitor bank should provide up to 100,000 KVAR.

From Eq. (5.6)

$$C = \frac{10^9 \cdot \text{KVAR}_c}{2 \cdot \pi \cdot f_c \cdot (kV_c)^2}$$

where kV_c = kilovolts applied across capacitors
  KVAR_c = reactive kilovolt-amperes delivered by capacitor bank

$$\text{KVAR}_c = 100 \cdot 10^3 \text{ kilovolt-amperes} \quad kV_c = 0.8 \cdot kV_M = 0.8 \cdot 284 = 227$$

The entire line-to-neutral voltage, 284 kV, is applied to the capacitor bank (R+L+C). The 0.8 factor is the estimated portion of the total line-to-neutral voltage drop during normal operation that occurs in the capacitors. In this 500-kV-line example, two or three (or more) capacitors in series will be required. The 0.8 factor reduces the estimated voltage drop in the stack of capacitors to 227 kV. The reader must keep in mind that the 0.8 factor impacts the capacitance required to obtain the desired 100 reactive megavolt-amperes per phase.

$$\frac{10^9 \cdot 100 \cdot 10^3}{377 \cdot 227^2} = 5.1476 \cdot 10^6 \quad \text{picofarads or 5.1476 microfarads}$$

Capacitor Bank Parameters per Phase

$$j = \sqrt{-1} \quad \varepsilon := 2.7182 \quad \omega := 377$$

$$R := 0.2 \text{ ohm} \quad L := 0.04 \text{ henry (H)} \quad C := 5.1476 \times 10^{-6} \text{ farad}$$

$$Z := \left[R + \left(\omega \times L - \frac{1}{\omega \times C} \right) j \right] = 0.2 - 500.2125j \quad |Z| = 500.2$$

$$\theta := \operatorname{atan}\left(\frac{-500.2125}{0.2}\right) = -1.57 \times \text{rad}$$

$$= -89.95 \text{ deg}$$

In a real design a step-down transformer might be used to simplify the design of the actual capacitor bank. But the transformer has inductance and that is not good. See Fig. 5-6. The capacitor bank installation changes the voltage at midpoint, but to simplify the solution of Example 5-1, we are assuming that Eq. (5.9) provides the midpoint line-to-neutral voltage before the ground fault and Eq. (5.10) provides the instantaneous line-to-neutral voltage.

Sinusoidal Instantaneous Midpoint Voltage and Current Flow before Ground Fault

Equation (5.9) provides the midpoint line-to-neutral voltage.

$$V_M = 217 \times 10^3 + 182\, j \times 10^3$$

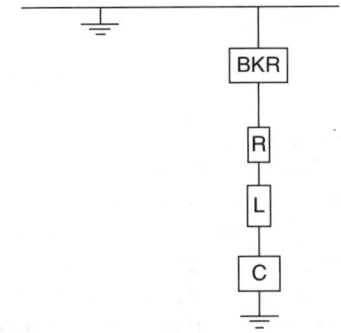

FIGURE 5-6 One phase of shunt capacitor bank short circuited by a line-to-ground fault.

Equation (5.10) provides the instantaneous midpoint line-to-neutral voltage.

$$v_M = \sqrt{2} \times 284 \times 10^3 \times \sin(\omega \times t + 0.698)$$

$$I_M = \frac{V_M}{Z} = \frac{217 \times 10^3 + 182 \, j \times 10^3}{0.2 - 500.2125 \, j} = -364 + 435 \, j$$

$$I_M = -364 + 435j \qquad |-364 + 435 \, j| = 567 \qquad (5.11)$$

$$\text{atan}\left(\frac{435}{-364}\right) = -0.874 \times \text{rad} = -50.08 \times \text{deg}$$

$$\pi - 0.874 = 2.268 \times \text{rad} \qquad 180 - 50.077 = 129.92 \text{ deg}$$

$129.92 - 40 = 89.92$ capacitor current phasor leads midpoint voltage by 89.92 deg

$i_M = \sqrt{2} \times 567 \times \sin(\omega \times t - 2.268)$ instantaneous capacitor current in capacitor branch

(5.12)

$V_c = Z_c \times I_M = \dfrac{-1}{\omega \times C} j \times (-364 + 435j)$ voltage drop across capacitor

$$\frac{-1}{\omega \times C} j \times (-364 + 435 \, j) = 224.1523 \times 10^3 + 187.5665 \, j \times 10^3$$

$$\text{atan}\left(\frac{187.5665}{224.1523}\right) = 0.697 \times \text{rad}$$

$$0.697 \, \text{rad} = 39.9 \times \text{deg}$$

$$V_c = 224.15 \times 10^3 + 187.57 \, j \times 10^3 \, j$$

$$|224.15 \times 10^3 + 187.57 \, j \times 10^3 \, j| = 292.28 \times 10^3$$

$v_c = \sqrt{2} \times (292.28 \times 10^3) \times \sin(\omega \times t + 0.697)$ instantaneous voltage across capacitor

Figure 5-6 depicts one phase of a neutral-grounded three-phase capacitor bank shunt connected to the power system. We assume that the two other capacitor phases are not affected by the line-to-ground fault or that their capacitors discharge very slowly. At t = 0, the instant at which the fault occurs, the voltage across the capacitor cannot change instantaneously and the current through the inductor cannot change instantaneously either. During a short circuit to ground conditions there is not a driving voltage source in the capacitor loop (see Fig. 5-6) and the circuit is not AC or DC. Furthermore, the transient will end when the energy stored in the capacitor is consumed in the resistance of the circuit. The discharge current prefers the copper path, but it does include the earth and the generator neutral connection to ground.

The differential equation that represents this discharge condition is

$$R \times i + L \times \left(\frac{d}{dt}i\right) + \frac{1}{C} \times \int i \, dt = 0 \qquad (5.13)$$

Differentiating Eq. (5.12), we obtain

$$R \times \left(\frac{d}{dt}i\right) + L \times \left(\frac{d^2}{dt^2}i\right) + \frac{i}{C} = 0 \qquad (5.14)$$

Equation (5.14) is a second-order linear differential equation representing the capacitor discharge into the specified R, L, and C circuit. Equation (5.15) below is the analytical solution of Eq. (5.14). The steady-state term of the analytical solution is zero because there is not a driving voltage.
When

$$\frac{R^2}{4 \times L^2} < \frac{1}{L \times C} \qquad \text{The transient term will be oscillatory, which is always the case if the capacitor is in the picofarad range.}$$

$$\frac{R^2}{4 \times L^2} = 6.25 \qquad \frac{1}{L \times C} = 4.857 \times 10^6$$

The values of the circuit parameters R, L, and C determine the frequency of the transient term which may be smaller, than equal to, or greater than 60 Hz.

$$i = \frac{-2 \times Q_0 \times \varepsilon^{-a \times t}}{\sqrt{4 \times L \times C - R^2 \times C^2}} \times \sin(\beta \times t) \qquad (5.15)$$

where Q_0 is the capacitor charge at $t = 0$ and

$$a := \frac{R}{2 \times L} = 2.5 \qquad \beta := \sqrt{\frac{1}{L \times C} - \frac{R^2}{4 \times L^2}} = 2203.775$$

The time to complete a cycle is the period T, and a cycle of Eq. (5.15) is completed when βT is equal to 2π radian. So

$$\beta \times T = 2 \times \pi \qquad T := \frac{2 \times \pi}{\beta} = 0.0029 = 2.9 \text{ milliseconds} \qquad (5.16)$$

The frequency of the oscillation in Eq. (5.15) is

$$f_{tr} = \frac{1}{T} = \frac{1}{2\times\pi} \times \sqrt{\frac{1}{L\times C} - \frac{R^2}{4\times L^2}} \quad f_{tr} = \frac{1}{T} \rightarrow f_{tr} = 350.742 \text{ cycles/second}$$

Assuming that the short circuit occurs when the midpoint voltage is at the maximum value, the peak of the sine curve, then

$$Q_0 := C\times(\sqrt{2}\times 284\times 10^3) = 2.07 \quad \text{initial capacitor charge, in coulombs}$$

$$\sqrt{4\times L\times C - R^2\times C^2} = 907.53\times 10^{-6}$$

$$i = \frac{-2\times 2.07\times \varepsilon^{-2.5\times t}}{907.53\times 10^{-6}} \times \sin 2203.8\times t$$

$T = 0.0029 \qquad 10\times 0.0029 = 0.029 \qquad 0.1 + 0.029 = 0.129$

$i = 4561.833\times \varepsilon^{-2.5\times t} \times \sin 2203.8\times t \qquad t := 0, 0.001..0.8$

Figure 5-7 is the plot of the capacitor bank discharge current during a single line-to-ground fault. In 800 milliseconds the discharge current goes from 4561 to 617 A. The discharge keeps going after the transmission line breakers isolate the fault because the line remains grounded. When a phase of the line fails to ground it could be, for instance, that

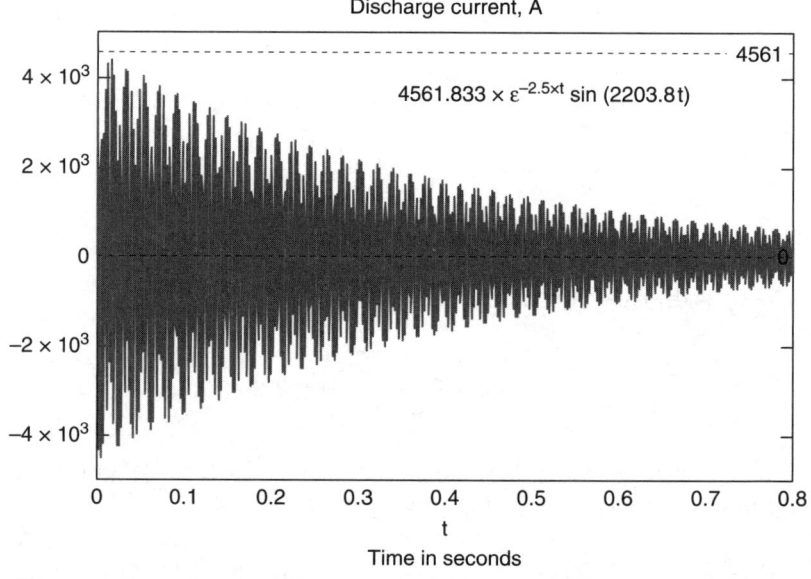

FIGURE 5-7 Capacitor bank discharge current during a single-line ground fault.

High-Voltage AC Capacitors

the line becomes connected to the metal enclosure of a transformer, which is always connected to the grounding grid. If the generator neutral is directly connected to the ground grid, then the capacitor discharge current will have a path of 100% metal. During 100 milliseconds, before the fault is cleared, one phase of the generator will conduct the short circuit current plus the no harmonic capacitor discharge current, which could be high or low frequency.

Capacitor Bank Charging Current Assuming It Is Completely Discharged

Assume the successful auto reclosing of the line breaker occurs 6 seconds after the second tripping that is after 2069 cycles of the discharge current [see Eq. (5.16)]. Six seconds is more than enough for the capacitor bank to be fully discharged, see Fig. 5-7.

The differential equation that represents the charging condition is

$$R \times i + L \times \left(\frac{d}{dt}i\right) + \frac{q}{C} = \sqrt{2} \times 284 \times 10^3 \times \sin(\omega \times t + 0.698) \quad (5.17)$$

where $\sqrt{2} \times 284 \times 10^3 \times \sin(\omega \times t + 0.698)$ is the instantaneous midpoint voltage applied to the bank. See Eq. (5.10). Differentiating Eq. (5.17), we obtain

$$R \times \left(\frac{d}{dt}i\right) + L \times \frac{d^2}{dt^2}i + \frac{1}{C} \times \left(\frac{d}{dt}q\right) = \sqrt{2} \times 284 \times 10^3 \times \omega \times \cos(\omega \times t + 0.698)$$

Substituting $d/dt\, q = i$ and dividing for L, we obtain Eq. (5.18).

$$\frac{d^2}{dt^2}i + \frac{R}{L} \times \left(\frac{d}{dt}i\right) + \frac{1}{L \times C} \times i = \frac{\sqrt{2} \times 284 \times 10^3 \times \omega}{L} \times \cos(\omega \times t + 0.698) \quad (5.18)$$

Equation (5.18) is a second-order linear differential equation representing the charging phase of the circuit depicted in Fig. 5-6 with the line fault to ground removed. The analytical solution of Eq. (5.18) would consist of two sinusoidal terms, a steady-state term, and a transient term. Instead, we will solve Eq. (5.18) by a numerical method.

Numerical Solution of Eq. (5.17)

A second derivative is a first derivative of a first derivative, so

$$\frac{d^2}{dt^2}[i(t)] = \frac{d}{dt}\left[\frac{d}{dt}[i(t)]\right]$$

Let us define two functions:

$$i_0(t) = i(t) \qquad i_1(t) = \frac{d}{dt}[i_0(t)]$$

Equation (5.18) can then be written as

$$\frac{d}{dt}[i_1(t)] + 5 \times i_1(t) + 4.857 \times 10^6 \times i_0(t) = 3785 \times 10^6 \times \cos(377 \times t + 0.698) \tag{5.19}$$

Equation (5.19) has two functions, $i_0(t)$ and $i_1(t)$, instead of one, $i(t)$. These two new functions are related by the following equation:

$$i_1(t) = \frac{d}{dt}[i_0(t)]$$

The original differential equation has been converted to a system of two differential equations:

$$\frac{d}{dt}[i_0(t)] = i_1(t) \tag{5.20}$$

$$\frac{d}{dt}[i_1(t)] = 3785 \times 10^6 \times \cos(377 \times t + 0.698) - 5 \times i_1(t) - 4.857 \times 10^6 \times i_0(t) \tag{5.21}$$

These two new equations had been written with the derivatives alone on the left-hand side of the equals sign. Next we define a single vector function, D, containing the two new functions of t in Eqs. (5.20) and (5.21) for the numerical solver. Vector D is the key for the numerical solution of differential equations.

$$D(t, i) = \begin{bmatrix} \frac{d}{dt}[i_0(t)] \\ \frac{d}{dt}[i_1(t)] \end{bmatrix}$$

$$D(t, i) := \begin{bmatrix} i_1(t) \\ 3785 \times 10^6 \times \cos(377 \times t + 0.698) - 5 \times i_1(t) - 4.857 \times 10^6 \times i_0(t) \end{bmatrix} \tag{5.22}$$

To write this vector, use MathCad array subscript to indicate the derivative order. We used the Rkadapt method to solve the differential equations. The Rkadapt(ic,t1,t2,npoints,D) returns a matrix of solution values for the differential equations specified by the derivatives in D and having initial conditions ic on the interval (t1,t2) using an adaptive step Runge-Kutta method. The initial values are evaluated at t1. At the instant of circuit energization, the capacitor bank current

High-Voltage AC Capacitors

is zero, the voltage across the capacitors is zero, and the charge is zero. Therefore, the initial conditions are all zeros.

Table 5-1 is the solution of differential Eq. (5.18), and provides the data for the plot shown in Fig. 5-8 that shows in great detail the capacitor bank charging current during the first 200 milliseconds.

Initial conditions for Table 5-1:

$$t1 := 0 \quad t2 := 0.2 \quad npoints := 800 \quad ic := \begin{pmatrix} 0 \\ 0 \end{pmatrix}$$

$$Q := Rkadapt(ic, t1, t2, npoints, D)$$

Initial conditions for Table 5-2:

$$t1 := 0 \quad t2 := 4.0 \quad npoints := 1200 \quad ic := \begin{pmatrix} 0 \\ 0 \end{pmatrix}$$

$$Q := Rkadapt(ic, t1, t2, npoints, D)$$

Table 5-2 is another solution of differential Eq. (5.18), and provides the data for the plot shown in Fig. 5-9 that shows in great detail the capacitor bank charging current during the first second. Table 5-2 has 1200 points and a time range of 4 seconds, so it provides a better

$$Q = $$

	t	$i_1(t)$	$\frac{d}{dt}(i_1)$
	0	1	2
0	0	0	0
1	0	0.015	6.952×10^5
2	0.001	0.113	1.327×10^6
3	0.001	0.368	1.89×10^6
4	0.001	0.839	2.378×10^6
5	0.001	1.57	2.789×10^6
6	0.002	2.589	3.117×10^6
7	0.002	3.909	3.36×10^6
8	0.002	5.525	3.517×10^6
9	0.002	7.415	3.584×10^6
10	0.003	9.541	3.563×10^6
11	0.003	11.848	3.453×10^6
12	0.003	14.262	3.254×10^6
13	0.003	16.698	2.97×10^6
14	0.004	19.052	2.602×10^6
15	0.004	21.21	...

TABLE 5-1 Solution Matrix of Eq. (5.18)

Chapter Five

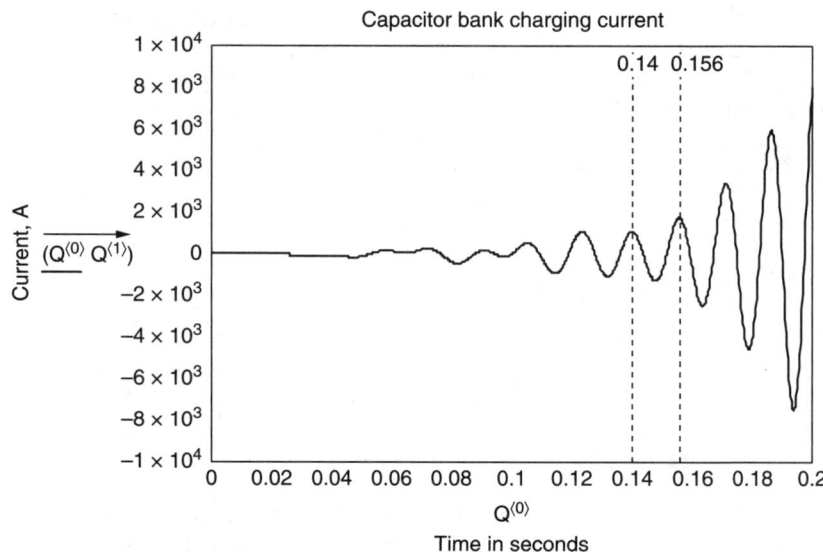

FIGURE 5-8 Capacitor bank charging current during the first 200 ms.

$$Q = \begin{array}{|c|c|c|c|} \hline & t & i_1(t) & \frac{d}{dt}(i_1) \\ \hline & 0 & 1 & 2 \\ \hline 0 & 0 & 0 & 0 \\ \hline 1 & 0.003 & 17.497 & 2.856 \times 10^6 \\ \hline 2 & 0.007 & -16.622 & -7.153 \times 10^6 \\ \hline 3 & 0.01 & -371.948 & -1.617 \times 10^7 \\ \hline 4 & 0.013 & -954.207 & -1.162 \times 10^7 \\ \hline 5 & 0.017 & -1218.43 & 4.337 \times 10^5 \\ \hline 6 & 0.02 & -1016.096 & 3.624 \times 10^6 \\ \hline 7 & 0.023 & -1064.434 & -6.042 \times 10^6 \\ \hline 8 & 0.027 & -1998.903 & -1.448 \times 10^7 \\ \hline 9 & 0.03 & -3211.489 & -8.837 \times 10^6 \\ \hline 10 & 0.033 & -3410.185 & 4.698 \times 10^6 \\ \hline 11 & 0.037 & -2445.734 & 9.23 \times 10^6 \\ \hline 12 & 0.04 & -1745.727 & 4.569 \times 10^5 \\ \hline 13 & 0.043 & -2345.239 & -7.262 \times 10^6 \\ \hline 14 & 0.047 & -3128.079 & -7.791 \times 10^5 \\ \hline 15 & 0.05 & -2092.281 & \ldots \\ \hline \end{array}$$

TABLE 5-2 Solution Matrix of Vector D in Eq. (5.21), 1200 Points

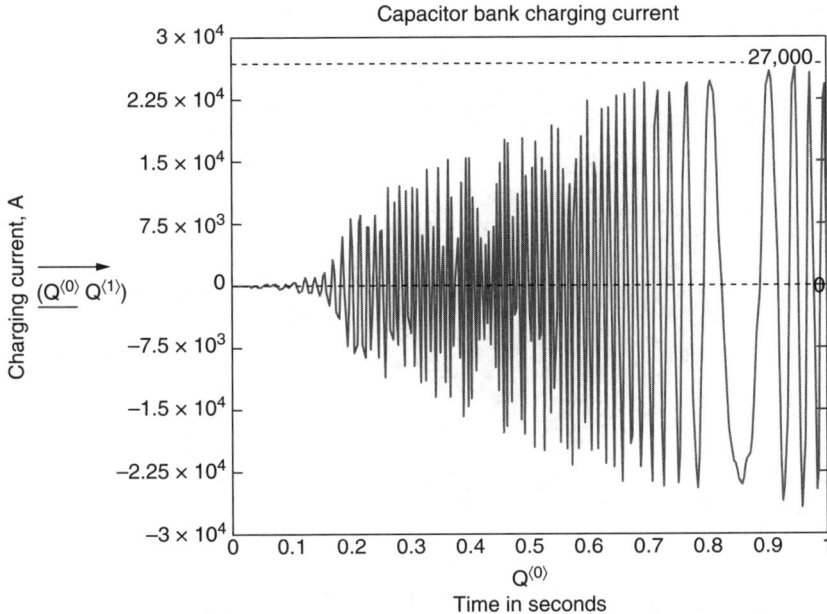

FIGURE 5-9 Capacitor bank charging current during the first second.

picture of the capacitor bank charging current during the transient condition.

Figures 5-9 and 5-10 show that there is a low-frequency component of approximately 1.18 Hz riding on top of the high-frequency component.

Computation of the Voltage Oscillations during the First Four Seconds

The impedance of the capacitor bank is

$$\left[R + \left(\omega \times L - \frac{1}{\omega \times C}\right)j\right] = 0.2 - 500.2125j \quad |Z| = 500.2$$

$$\theta := \operatorname{atan}\left(\frac{-500.21}{0.2}\right) = -1.57 \times \text{rad} = -90 \text{ deg}$$

The value of the peak of the voltage oscillation is

$27{,}000 \times 500.2 = 13.5054 \times 10^6$ volts or 13505 kV in a 500 kV line

$\dfrac{500}{\sqrt{3}} = 288.675$ rated line-to-neutral voltage, kV

$\dfrac{13{,}505}{288.7} = 46.779$

124 Chapter Five

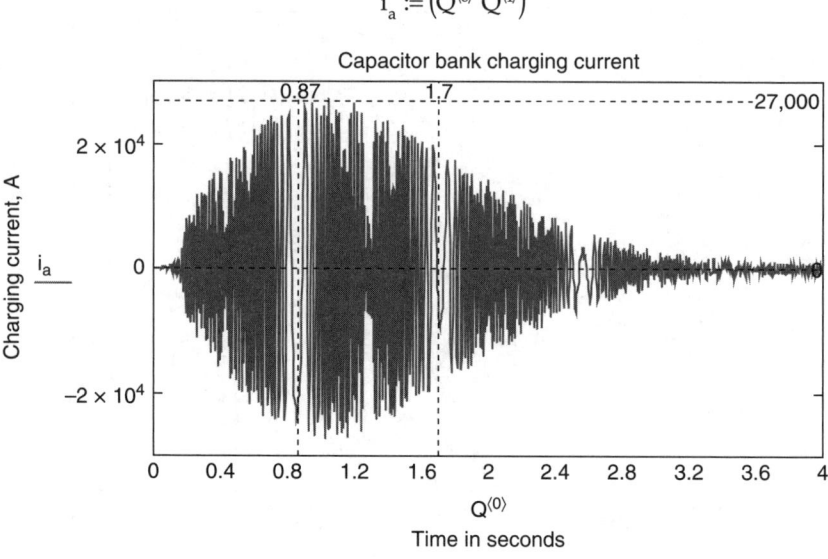

FIGURE 5-10 Capacitor bank charging current during the first 4 seconds.

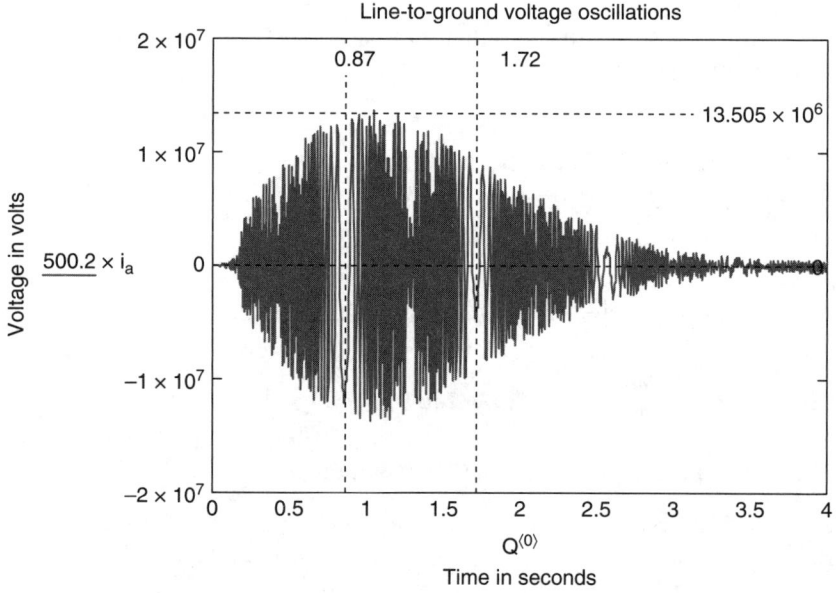

FIGURE 5-11 Line-to-neutral voltage oscillations produced during charging the capacitor bank.

The peak of the voltage oscillation is more than 46 times the rated voltage. This extremely high voltage, higher than 13 million volts, will produce flashovers on buses and transmission line insulators and over equipment bushings. Furthermore, it will cut through equipment insulation, such as transformers and breakers.

Figure 5-11 shows high- and low-frequency voltage oscillations. The low frequency, 2.273 Hz, is clearly nonharmonic.

$$T = 1.72 - 0.87 = 0.85 \text{ s} \qquad f = \frac{1}{T} = \frac{1}{0.85} = 1.18 \text{ cycles/second}$$

This nonharmonic low-frequency oscillation induces intense vibrations and noise in synchronous machines. It could be very damaging for any generator connected to the transmission line if their rotors' natural frequency of oscillation is close to the capacitor-induced low-frequency oscillation.

CHAPTER 6
Substation Grounding

6.1 Background

In the construction of alternating-current (AC) electrical substations, it has been a long-standing practice to install a grid of buried metal conductors over the entire substation yard and to solidly connect this grid to the neutral of the AC power system. The purpose of this buried grid is to make the substation yard a safe place even when a fault to ground occurs, and to provide a path of low impedance between the neutral of the power system and earth. A good connection between the neutral of the power system and earth is vital to maintain the voltage gradient between them as low as possible under all operating conditions. Good neutral grounding enhances the substation safety, helps to stabilize the output voltage during transient condition, dissipates the energy produced by lightning strokes, and improves the performance of the electrical apparatus connected to the power system. In addition, within a modern substation there are installed a multitude of electric power, control, instrumentation, and communication systems. For all these systems to operate satisfactorily, it is necessary to minimize the electromagnetic interference among them. Unfortunately, all have a common point—the ground—the natural sink where all the spurious interference and noise contaminate all the systems connected to ground. Although it is not a complete cure, it is better to have separate grounding systems for high-frequency equipment, especially if switching devices are involved.

6.2 Approaches to Grid Design

The best way to make a substation yard a safe place for any person working or walking inside or just outside the perimeter of the substation is to keep the voltage gradient that he or she may experience below the tolerable touch, step, and transferred voltages. The grid also must provide an easy path for the return current (zero sequence) when a short circuit to ground occurs. This zero sequence current must return to the neutral of the generator (or generators) feeding the fault. There is no return current when a three-phase symmetrical fault occurs. In this case the potential of the point at which the fault occurs is the same as the potential of the neutral point of the generator. The discharge of lightning does not produce a return current. Lightning is an electrostatic phenomenon produced by electric charges that have become separated and organized into plus and minus charges. The separated and organized charges produce an electric field between them. When the field becomes strong enough to overcome the dielectric strength of the air separating the charges, this produces the lightning which neutralizes the polarized charges (the discharge ends the charge separation). The generally accepted criterion used in grounding mat design is the *tolerable touch voltage*. However, in every design a final check is made of the actual step and transferred voltages to be sure that they are smaller than the tolerable touch voltage. Chapter 7 covers the injuries inflicted on a human by the passage of an electric current. In Chap. 8 the grid design procedure used takes care of the effects of nonuniform leakage current distribution by means of experimentally determined correction factors.

In Chap. 9 the earth is considered homogeneous; specifically the earth resistivity is assumed constant. A method is given to determine the actual leakage current distribution during a fault to ground. Also covered is the computation of the surface potential at any point in the substation yard. Chapter 10 provides a better representation of the actual grounding system. In it the earth is represented by a two-layer model. It is impractical to use earth models of more than two layers, because the computations become very complex and the large amount of reliable field data and how they changes with the seasons and weather is never available.

6.3 Generally Accepted Assumptions

The following statements are assumed to be true and are commonly used in electrical substation grounding system design.

- The resistance of the human body is considered to be 1000 ohms (Ω), regardless of the current path.
- The tolerable body current is shown in Fig. 7-1 as a function of the duration of the electrical shock and is given by Eq. (7.1).
- The contact resistance between a human foot and earth is given by Eq. (7.4) and is equal to 3 times the resistivity of the upper soil layer.
- For any accident in which the current flow path is from hand to feet, the combined resistance of the two feet in parallel is 1½ times the resistivity of the upper soil layer.

6.4 Separated Ground Rods

High-frequency electronics are present in practically all substations. Among the many apparatuses are high-frequency signal carriers, thyristor switched capacitor banks, and thyristor switched reactors.

These apparatuses should have dedicated and separate, grounding rods. Lightning arresters, shield wires, and shield masts should also have separate ground rods.

As the cost of land for air-insulated substations increases, or in the case of some metropolitan areas in which the required space is simply not available, the use of gas insulate substations (GIS) is increasing. The area covered by a GIS is much smaller than the area within the fence of an air insulated substation. Therefore, GIS depend more on ground rods for their earthen requirements than an air-insulated substation would.

6.5 Substation Fences

The reader must be aware that in this subject the legal standard is the latest revision of the *IEEE Guide for Safety in AC Substation Grounding*, IEEE Standard 80.

The grounding of substation perimeter fences is an important subject because the perimeter fence is accessible to the general public and should be safe for anyone to touch or walk around. So the fence grounding designer must guarantee that the touch and step potentials are within the tolerable limits established by all applicable standards. A metal fence should, in general, be continuous with no gaps or isolated spans. And it must be bonded to the grid at regular intervals, such as at every pole. The design of the fence grounding system can be approached in two different ways:

1. The substation grounding grid or mat is extended at least 1.5 meters (m) beyond the fence perimeter. If this is the design approach taken, then most of the mat grounding rods should be buried and connected to the grid at the perimeter conductor.

2. The fence has its own separate grounding system that is not intentionally coupled to the main grounding grid of the substation. The separated grounding grid should consist of mat and rods.

Approach 1 has potentially the higher cost; it requires more copper and more land. But it reduces the total resistance to earth of the grid. Approach 2 type fences must be isolated from the substation main grounding grid, and this, with time, may be hard to accomplish. The sneaky interconnection may be a water pipe, a conduit, a cable, anything that spans the separation gap. And it might occur underground or by air.

Another kind of problem is presented by gates, especially equipment gates. Swing doors installed in approach 1 fences should be swing-in, not swing-out.

Reactive power compensation equipment with air core reactors located near to the fence might induce circulating currents in the fence. If this is the case, and you can't move the reactor, you would need to divide the fence in many sections electrically isolated from each other.

CHAPTER 7
Dangerous Electric Currents

7.1 Background

The safe maximum electric current flow through the human body depends on the kind of electric current, the physical condition of the person receiving the electric shock, and the phase of the heart cycle at the instant that the shock occurs. Humans can tolerate much higher values of direct current than 60-hertz (Hz) alternating current, and even higher values of extremely short pulses of current, such as the ones produced by lightning strikes.

The threshold at which ventricular fibrillation of the heart occurs is higher at 25 Hz than at 60 Hz and even higher for DC shock, provided that the duration of the shock is at least 1 second. The literature surveyed does not go into any detail regarding the physical condition of the person receiving the electric shock other than assuming that he or she is a "normal and healthy" individual. The human body resistance to an electric shock depends on many factors, such as the magnitude of the voltage applied, current flow path, size and weight of the victim, skin condition, kind of shoe the victim is wearing, and prevailing atmospheric humidity. The factors considered in this chapter are electric current magnitude, frequency, duration of the electric shock, and current flow path.

Many of the present ideas about the response of the human body to an electric shock are based on tests conducted at Columbia University with animals. The results of these tests were reported is May 1936 by Ferris, King, Spence, and Williams. Based on these tests, Dalziel derived an empirical equation that gives the tolerable body current as a function of time; the Dalziel equation is widely used in engineering practice. It must be understood that in this subject the available data are often incomplete, inconsistent, or qualitative in nature.

7.2 Magnitude and Frequency

The effects of injuries produced in humans by electric shocks are classified as follows:

1. *Perception.* At 60 Hz the threshold of perception is from 0.5 to 1.5 milliamperes (mA) for men and from 0.2 to 1.0 mA for women. For frequencies larger than 300 Hz the perception current threshold increases with frequency. For instance, at 70 kilohertz the perception current for men is 100 mA.

2. *Reaction or surprise.* Electric shock currents between 1 and 3 mA (for men) might provoke a sudden involuntary reaction in the victim which could induce an accident.

3. *Loss of muscular control.* It happens for electric shock currents between 10 and 22 mA for men and from 6 to 14 mA for women. For shock currents of these magnitudes the victim is frozen by a severe spasm of the hand muscle, and she or he is not capable of releasing an energized object grasped with his or her hand. Dalziel defines the *let-go current* as "the maximum current an individual can tolerate and still be able to release his grasp of an energized conductor by using muscles directly stimulated by that current."

4. *Muscular inhibition.* Electric shock current larger than the let-go value, but smaller than the value that causes ventricular fibrillation of the heart, produces a severe contraction of the chest muscles and makes breathing difficult or impossible. At this level of current the victim still has a chance for recuperation if the current is interrupted within a few seconds; however, the victim could die, if the current is allowed to flow for longer periods.

5. *Ventricular fibrillation of the heart.* The threshold of ventricular fibrillation, at 60 Hz, occurs in the range of 50 to 100 mA. Fibrillation is followed by ventricular standstill. Actually, once the heart of the victim starts to fibrillate, death usually occurs within minutes. It is the general consensus that currents of much higher magnitude can be tolerated if the duration of the electric shock is very short. The threshold of ventricular fibrillation is higher at 25 Hz than at 60 Hz, and even higher for a direct current shock, provided that the duration of the shock is at least one second.

6. *Severe electric shock.* When the magnitude of the electric shock is large enough, the victim could experience some of or all the following symptoms: respiratory paralysis, unconsciousness, and severe burn. In this eventuality death is the usual outcome.

7.3 Duration and Current Path

One hundred milliamperes is the round number used in electrical engineering practice to designate the magnitude of the 60-Hz electric current that produces ventricular fibrillation, which is one of the leading causes of death by accidental electrocution. It is known that many other factors besides magnitude affect the actual threshold of ventricular fibrillation. Dalziel used 110 pounds (lb) as the victim's average weight (to be safe) when he derived his empirical equation. The Dalziel equation gives the maximum current not likely to produce ventricular fibrillation as a function of time. Actually, he predicted that 99.5 percent of the victims will not experience ventricular fibrillation if the current is given by Eq. (7.1). The graph shown in Fig. 7-1 is a plot of the tolerable body current as given by Eq. (7.1).

$$I_b = \frac{0.116}{\sqrt{t}} \tag{7.1}$$

where I_b = rms value of current through human body, amperes
        t = duration of electric shock, seconds
        0.116 = empirical constant

Tolerable body current plot:

$$t := 0.01, 0.015..2.00$$

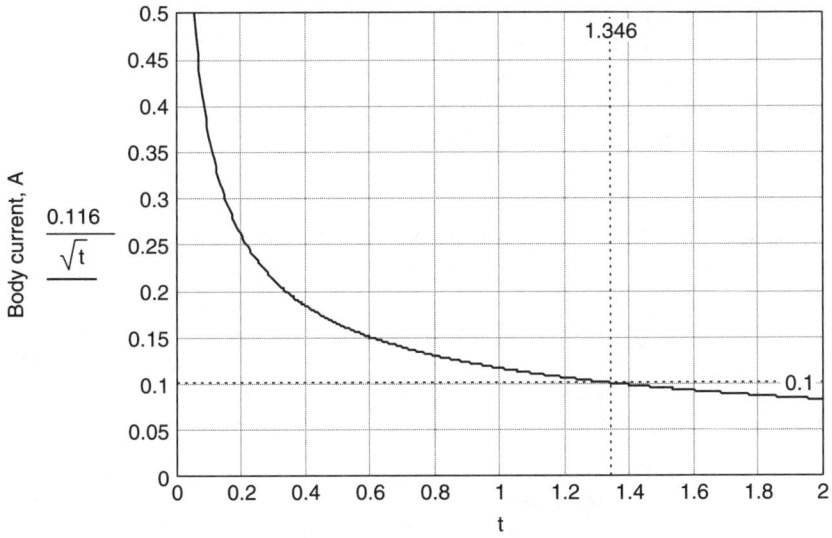

Figure 7-1 Tolerable body current.

If ventricular fibrillation occurs at 100 mA, then

$$\sqrt{t} = \frac{0.116}{0.100} = 1.16 \quad 1.16^2 = 1.346 \quad \text{time to fibrillation}$$

$$= 1.346 \text{ seconds}$$

Equation (7.1) is based on a shock duration of 0.03 to 3 seconds, 60-Hz alternating current, and a current flow path from hand to feet. This is the most probable path for accidental shock, and it is also one of the more dangerous. Another path that is important in the design of substation grounding systems is from one foot to the other. The tolerable foot-to-foot current is larger than the value given by Eq. (7.1). Because of that, the current value given by Eq. (7.1) is the criterion used when designing grounding grids, regardless of the current flow path. The maximum potential difference that a person can withstand in an electric shock accident, in which the body current flows from foot to foot, is called the *tolerable step voltage*. Multiplying the tolerable body current, as given in Eq. (7.1), by the resistance of the foot-to-foot path, the tolerable step voltage is obtained.

$$E_{stap} = (R_b + 2 \cdot R_F) \cdot \frac{0.116}{\sqrt{t}} \quad \text{volts} \tag{7.2}$$

where R_b = resistance of human body; 1000 ohms is accepted value regardless of current path
R_F = contact resistance of a foot, ohms
t = duration of electric shock, seconds

To determine R_F, the foot is replaced by an equivalent circular plate electrode implanted on the surface of the earth; the radius of the equivalent plate is 0.0833 meters. The resistance to earth of a circular plate electrode is given by the Laurent's formula as:

$$R_F = 0.25 \cdot \frac{\rho_s}{r} \quad \text{ohms} \tag{7.3}$$

where ρ_s is the resistivity of the soil near the surface, in ohm-meters, and r is the radius of the equivalent circular plate, in meters.

$$R_F = \frac{0.25}{0.0833} \cdot \rho_s = 3 \cdot \rho_s \tag{7.4}$$

Substituting in Eq. (7.2), we obtain the following expression for the tolerable step voltage:

$$E_{step} = (1000 + 6 \cdot \rho_s) \frac{0.116}{\sqrt{t}}$$

$$E_{step} = \frac{116 + 0.7 \cdot \rho_s}{\sqrt{t}} \quad \text{volts} \quad \text{tolerable step voltage} \tag{7.5}$$

The maximum potential difference that a person can withstand in an electric shock accident, in which the current flows from a hand to both feet, is called *tolerable touch voltage*. Multiplying the tolerable body current by the resistance of the hand-to-feet path, we obtain the tolerable touch voltage.

$$E_{touch} = \left(R_b + \frac{R_F}{2}\right) \cdot \frac{0.116}{\sqrt{t}} \quad \text{volts} \tag{7.6}$$

$$E_{touch} = (1000 + 1.5 \cdot \rho_s) \frac{0.116}{\sqrt{t}}$$

$$E_{touch} = \frac{116 + 0.17 \cdot \rho_s}{\sqrt{t}} \quad \text{volts} \quad \text{tolerable touch voltage} \tag{7.7}$$

Figure 7-2 shows the tolerable step voltage for several values of the soil resistivity near the surface in ohm-meters.

Figure 7-3 shows the tolerable touch voltage for several values of ρ_s.

Another important type of touch voltage is the *let-go voltage*, defined as

$$E_{let-go} = (1000 + 1.5 \cdot \rho_s) \cdot 0.009 = 9 + 0.0135 \cdot \rho_s \quad \text{volt} \tag{7.8}$$

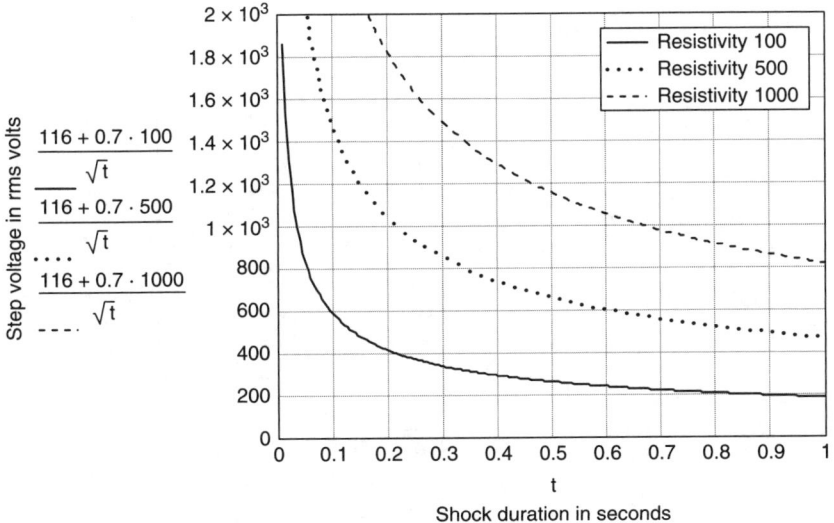

FIGURE 7-2 Tolerable step voltage.

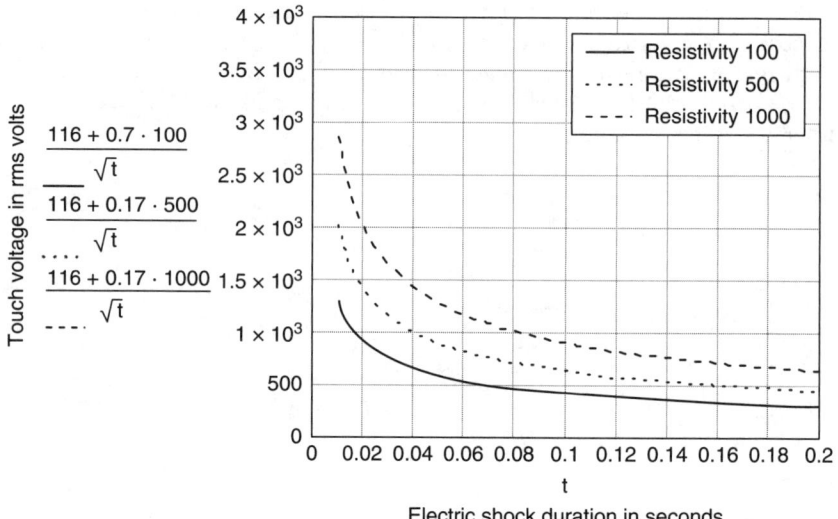

Figure 7-3 Tolerable touch voltage.

where ρ_s is the soil resistivity and the let-go current is the maximum current an individual can tolerate and still be able to release his or her grasp of an energized conductor by using muscles directly stimulated by that current. The accepted value of the let-go current at 60 Hz is 9 mA. Figure 7-4 shows the let-go voltage for several values of ρ_s.

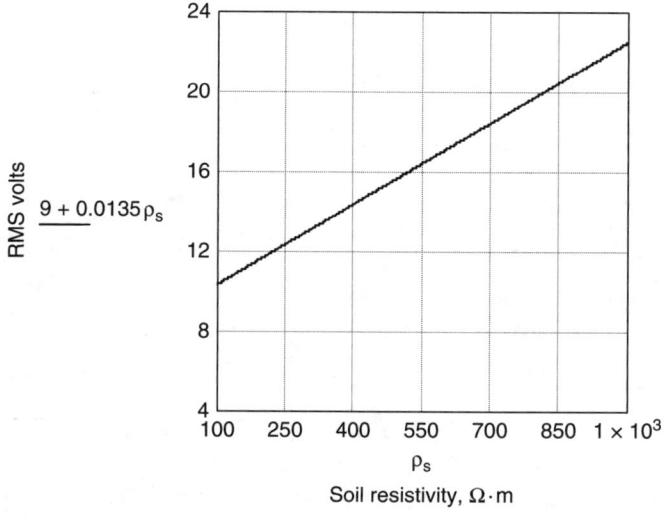

Figure 7-4 Let-go voltage.

The following assumptions were made to deduce Eqs. (7.5) through (7.8).

1. At 60 Hz the human body and earth soil can be considered as pure resistive circuit elements.
2. The mutual resistance between feet can be neglected.
3. Hand contact resistance and shoes' resistance were neglected.
4. The victim's skin was not damaged by the electric shock; punctured and burned skin may reduce the body resistance below 1000 ohms.
5. The electric shock duration falls in the range of 0.03 to 3 seconds.
6. Homogeneous soil with constant resistivity.

It should be stated that Eqs. (7.5) and (7.7) were deduced without considering the length of the step or arm-reach of the victim. These factors are not required because the effect of the mutual resistance is neglected.

7.4 Electrical Substation Grounding

The design of electrical substations usually includes a grounding grid of metal conductors or grounding mats that frequently covers the entire area of the substation. The grid is buried shallow (a meter deep approximately), and it is connected to all non-current-carrying metal objects on the substation yard, such as transformer tanks, breaker enclosures, and supporting metal structures. The grid is also connected to some points of the electrical system, such as transformer neutrals, neutral grounding devices, and lightning arresters. During normal mode of operation the grid and all the interconnected metal objects are essentially at the same potential as the earth surface within the substation.

When a short circuit to ground (non-current-carrying metal objects) occurs within the substation, all the grid conductors and interconnected metal objects become energized at the same potential (neglecting the voltage drop in the metal conductors). The magnitude of this potential is a fraction of the affected electrical system line-to-neutral voltage. And the increase in potential of the earth surface directly above the ground grid is smaller than the potential increase of the metal conductors of the ground grid. In fact, a potential gradient is established between the ground grid and the earth surface. Furthermore, due to variations in the earth current density, the earth surface is not an equipotential surface during short circuit conditions. So a person walking is the substation yard or touching any of the grounded metal objects will suffer an electric shock, produced by the difference in potential between the two points on the earth surface spanned by his or her legs or by the difference in potential between the point on the earth surface where the person is standing and the metal object touched by his or her hand.

For a careless person who touches a live conductor during normal mode of operation, the grounding grid does not offer any protection at all. He or she is actually submitted to the full line-to-neutral voltage of the electrical system.

7.5 Important Voltage Gradient Definitions

At this point is convenient to summarize the definitions of step and touch voltages and to introduce the definition of mesh voltage.

Tolerable step voltage (E_{step}) is the maximum potential difference that a person can withstand when the body current path is from foot to foot. See Eq. (7.5).

$$E_{step} = \frac{116 + 0.7 \cdot \rho_s}{\sqrt{t}} \quad \text{volts}$$

Step voltage (V_{step}) is the actual potential difference between two points on the substation surface that are separated by 1 meter.

Tolerable touch voltage (E_{touch}) is the maximum potential difference that a person can withstand when the body current path is from either hand to both feet. See Eq. (7.7).

$$E_{touch} = \frac{116 + 0.17 \cdot \rho_s}{\sqrt{t}} \quad \text{volts}$$

Touch voltage (V_{touch}) is the potential difference between the place on the earth surface where a person could be standing and any grounded object that can be touched with either of the person hands.

Mesh voltage (V_{mesh}) is the maximum potential difference between the mesh conductor and a point of the earth surface within the mesh.

Transferred voltage (V_{trans}) is the name assigned to touch or step voltages, inside the substation when the voltage gradient originates outside the substation. Also the transferred voltage could originate inside the substation and then be exported to remote places outside the substation. The means for this to happen are buried pipes, underground cables, shield wires, etc.

CHAPTER 8
Ground Grid Preliminary Design

8.1 Background

To make the substation yard a safe place for human, established design practice is to keep the voltage gradients that a person may experience below the tolerable touch and step voltages. This gradient control is accomplished by a grid of buried metal conductors over the entire substation yard. Often, the grid is extended beyond the substation fence 1 meter (m) or more. The grid consists of bare copper (rarely is other metal used) cables buried horizontally and shallow to form a network of square or rectangle meshes with cables bonded together at each intersection. Vertically oriented grounding rods are frequently connected at some of the cable intersections, especially at the corner meshes. Figure 8-1 is an illustration of a typical grounding mat. Once the soil resistivity measurements of the natural soil are completed, the interconnection requirements specified, and the equipment and structure's layout known, the grid designer could go ahead with the preliminary design of the substation grounding system. She or he needs to determine the following:

1. The resistance to earth of grounding electrode system.
2. The rms value of the worst ground fault current.
3. The grid's conductor size.
4. The number of conductors and spacing among them.
5. Orientation of the grid conductors. The grid conductors should be installed parallel to the rows of equipment and structures.

The first four items of the above list are interrelated, and none can be determined separately from the others. The practice is to use an iterative design procedure in which repetition of the computations produces better results.

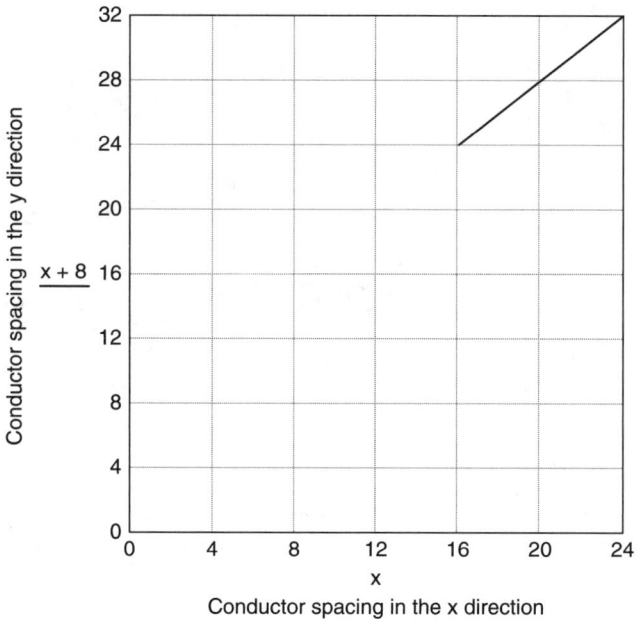

Figure 8-1 Typical ground mat.

Mat area: 24 × 32 = 768 m²
Conductor size: 2/0 copper cable
Conductors spacing: 4 meter
Depth of burial: 1 meter
Soil resistivity: 200 ohm-meters ($\Omega \cdot m$)
Four vertically installed copper-clad steel rods, one in each corner mesh, 0.0254 meter in diameter and 6.1 meter long. The location of the maximum touch voltage probably occurs along the diagonal of any of the corner meshes, as indicated in Fig. 8-1. The maximum step voltage also occurs along the same line but just outside the mat.

8.2 Single-Rod Electrodes

The most practical and most used grounding electrode is the ground rod. So let us discuss the equation for the resistance to earth of a single rod vertically buried in homogeneous soil, with the top flush with the surface of the earth. If the current distribution along the rod is uniform and the length of the rod is much larger than its diameter, then the rod resistance to remote earth is given by Eq. (8.1).

$$R_1 = \frac{\rho}{2\pi \times L} \times \ln\frac{4 \times L}{d} \quad \text{ohms} \tag{8.1}$$

where ρ = resistivity of homogeneous soil, ohm-meters (Ω · m)
L = length of rod, meter
d = rod diameter, meter

From Eq. (8.1) it is obvious that to decrease the resistance to earth, increasing the length is more effective than increasing the diameter. The resistance to earth of a vertical rod is affected by the length and the logarithm of the length; however, the rod's diameter enters Eq. (8.1) in the argument of a logarithm only. Figure 8-2 shows the resistance to earth of a single vertical rod in homogeneous soil as a function of its length. And Fig. 8-3 shows the rod's resistance to earth as a function of its diameter.

$\rho := 200$ ohm meter $\quad d := 0.0254$ meter $\quad L := 1, 2 .. 10$ meter

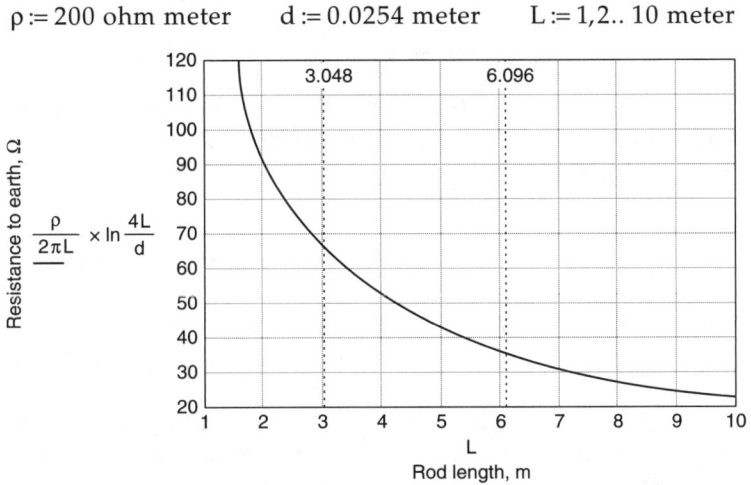

FIGURE 8-2 Single rod's resistance to the earth as a function of its length.

$\rho := 200$ ohm meter $\quad L := 6.1$ meter $\quad d := 0.0127, 0.01905, 0.0635$

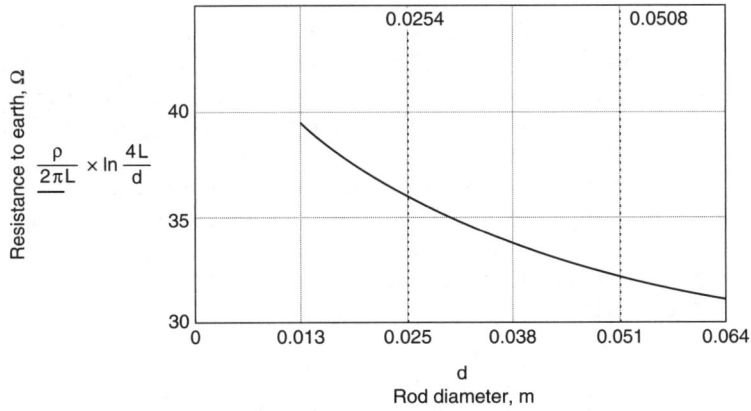

FIGURE 8-3 Rod's resistance to earth as a function of its diameter.

8.3 Ground Mat Resistance to Earth, Approximated Formulas

The resistance to earth of a ground mat could be calculated using Eq. (8.2), which is used extensively even though it provides only approximate results.

$$R = \frac{\rho}{4 \times r} + \frac{\rho}{L} \quad \text{ohms} \tag{8.2}$$

where ρ = average earth resistivity, ohm meter
  r = radius of a metal disk with same area as that covered by mat, meter
  L = total length of buried grounding conductors, meter

Equation (8.2), introduced by P. Laurent in 1951, is still very useful and frequently applied to the design of grounding mats for the following reasons.

1. It is very simple to use.
2. The computation of the resistance to earth of ground mats is never of high accuracy, because the data available are frequently incomplete or nonexistent.
3. The earth soil is not homogeneous, and its resistivity changes both vertically and horizontally. In fact, several layers of soil with different resistivity are to be expected in most substation sites. The layers' depth changes within the construction site, and therefore they are neither perfectly horizontal nor parallel to the mat.
4. The soil layers' resistivity changes with the seasons, because of temperature and soil moisture content variations.
5. The mat's copper cables are laid out on the boundary line between natural soil and backfill material.

The term $\rho/4r$ gives the resistance to earth of a metal disk near the soil surface with the same area as the ground mat. Because the mat resistance must be larger than the resistance of the solid metal disk, to compensate, Laurent added the term ρ/L to his equation. The magnitude of this term decreases as L increases.

For a given substation site, the resistance to earth of the ground mat is principally determined by the area covered by the ground mat. The specific mat shape is of secondary importance. Another approximated equation that provides the resistance to earth of a ground mat is deduced as follows:

$$A = \pi \times r^2 \qquad r := \sqrt{\frac{A}{\pi}}$$

where A is the area of the ground mat and r is the radius of the metal disk with the same area as the ground mat.

$$R = \frac{\rho}{4 \times r} \quad (8.3)$$

where R is the resistance to earth, near the soil surface, of a metal disk of radius r.

$$R = \frac{\rho}{4 \times \sqrt{A/\pi}} = \frac{\rho \times \sqrt{\pi}}{4 \times \sqrt{A}} \quad (8.4)$$

Equation (8.4) provides the resistance to earth of a solid metal disk with the same area as the ground mat. Note than the mat is a network of metal cables, not a solid plate.

Example 8-1
Compute the resistance to remote earth of the ground mat and of each single rod for the typical ground mat of Fig. 8-1.
Single-rod resistance:

$$R_1 = \frac{\rho}{2\pi \times L} \times \ln \frac{4 \times L}{d} = \frac{200}{2 \times \pi \times 6.1} \times \ln \frac{4 \times 6.1}{0.0254} = 35.836 \text{ ohms}$$

$$A := 768 \qquad \pi \times r^2 = 768 \qquad r := \sqrt{\frac{768}{\pi}} \qquad r = 15.635$$

$$L := 9 \times 24 + 7 \times 32 = 440$$

Grounding mat resistance:

$$R = \frac{\rho}{4 \times r} + \frac{\rho}{L} = \frac{200}{4 \times r} + \frac{200}{440} = 3.652 \text{ ohms} \qquad \frac{200}{4 \times r} = 3.198$$

The resistance to earth of a well-designed mat should be between 1 and 5 ohms.
We have similar results applying Eq. (8.4):

$$R = \frac{\rho \times \sqrt{\pi}}{4 \times \sqrt{A}} = \frac{200 \times \sqrt{\pi}}{4 \times \sqrt{768}} = 3.198 \text{ ohms}$$

The 3.198-ohms result is equal to the first term of the ground mat resistance.

8.4 Ground Mat Conductor Corrosion

Different metals in the presence of a soil electrolyte (wet soil) constitute a galvanic cell. In normal operating conditions, potential differences between metals produce an electric current between them,

Metal	Ion	E^0, Volt	
Aluminum	Al^{3+}	−1.662	Anodic
Zinc	Zn^{2+}	−0.763	Least noble
Iron	Fe^{2+}	−0.440	Corroded
Tin	Sn^{2+}	−0.136	
Lead	Pb^{2+}	−0.126	
Hydrogen	H^+	0.000	Reference
Copper	Cu^{2+}	+0.337	Cathodic
Copper	Cu^+	+0.521	Most noble
Silver	Ag^+	+0.799	Reduced

TABLE 8-1 Electrochemical Potential Series at 25°C

provided there is a path for the flow of current from one to the other. The result is corrosion of the metal with lower potential (least noble). The potential difference of the galvanic cell (electromotive force) and the resulting electrochemical corrosion depends on the position of the metals in the electrochemical potential series. Table 8-1 provides a list of the electrochemical potentials of some common metals that might be buried in the soil of a substation yard; E^0 is the standard electrode potential at 25°C for cells with all reactants and products at unity activity. A standard cell contains ions at 1 mole concentration and gases at 1 atmosphere pressure.

The potentials listed in Table 8-1 are referenced to the standard hydrogen electrode; an arbitrary value of 0 volts is assigned to the reference electrode. The maximum electromotive force that a galvanic cell can develop is the difference between the two electrode potentials as listed in Table 8-1.

Metals with higher electrochemical potentials are listed at the bottom of Table 8-1; they are positive, noble, or stable and they are called *cathodic*. Metals with lower electrochemical potentials are listed at the top of the table; they are negative, less noble, or stable and they are called *anodic*. These metals at the top of the table constitute the oxidized side of the reaction; they would corrode.

Grounding mats are made, in most cases, with bare stranded copper cables which are corrosive to practically any other metal buried in the substation site. Copper cables are the most common choice in the design of grounding grids for these reasons:

- High electrical conductivity.
- It does not corrode in most soils.

- It is cathodic with reference to most any other metal that might be buried in the substation yard, and therefore will not corrode because of their presence.

However, the use of copper cable has the following drawbacks:

- Copper corrodes in soils containing salts or compounds of ammonia, and sulfur.
- It corrodes metals with negative electrochemical potentials; for instance, it accelerates the corrosion of buried steel pipes, anchor assemblies, metal foundations, and well metal casings.
- It encourages stealing, because it is easy to sell, at a good price, as scrap metal, especially during the installation period.

Underground corrosion control is an important factor in the selection of the substation site and the metal used in the ground mat.

8.5 Grid Conductor Size

Substation grounding systems must last in good operating condition for an indefinite period of time. The entire installation including grid conductors, ground rods, equipment ground leads, and conductor joints and splices must be mechanically rugged and capable of resisting the soil corrosive attack and the high currents and temperatures resulting from electrical system faults to ground. When a fault to ground occurs, the temperature of the ground mat rises, resulting in the reduction of some of its mechanical properties and in accelerated corrosion. Fortunately, these effects are inconsequential, because ground faults last only a short time. The critical thing to avoid is the fusing or melting of equipment ground leads, grid conductors, conductor joints, and splices (conductor splicing must be avoided). The formula for calculating the minimum size of grounding copper conductors to avoid fusing is

$$\left(\frac{A}{I}\right)_{copper} = \sqrt{\frac{33 \cdot t}{\log[(T_m + 234) \div (T_a + 234)]}} \quad \text{circular mils/ampere} \quad (8.5)$$

where A = copper conductor cross-sectional area, circular mils
  I = short circuit current, amperes (A)
  t = duration of short circuit current flow, seconds
  T_m = maximum allowed temperature, in degrees Celsius (°C)
  T_a = conductor temperature before short circuit (°C)

Equation (8.5) is based on the assumption that the heat loss is negligible, in other words that all the energy produced by the short circuit current is kept within the conductor and is effective in raising its temperature. This assumption is valid provided that the fault duration is very short, less than 1 second. The maximum allowed temperature for copper conductors is 1083°C, which is the melting point of copper. However, the maximum allowed temperatures for copper cable joints are as follows:

Exothermic joints	1083°C (same as conductor)
Brazed joints	450°C
Bolted joints	250°C (not recommended)

The value of T_m should be selected in accordance with the specific design, but in almost every case, substation ground mats are installed using exothermic connections. Therefore, from here on T_m will be considered equal to 1083°C. Before a fault to ground, the temperature of the ground mat conductors is considered to be equal to the ambient temperature, or $T_a = 40°C$. This is the appropriate assumption if there is not automatic reclosing of the circuit breaker that protects the faulty circuit. Breaker reclosing is a common mode of operation in utility company substations. The initial grid conductor temperature immediately after the fault is 40°C; however, the computation of the initial temperature rise before the first reclosing operation and the computation of the grid conductor temperature rise after each reclosing operation are a complex problem that requires one to know the thermal conductivity and thermal time constant of the substation yard soil. This subject is beyond the scope of this book.

A typical reclosing sequence is as follows:

First reclosing occurs 0.1 seconds after first tripping.
Second reclosing occurs 3 seconds after second tripping.
Third reclosing occurs 30 seconds after third tripping.
Because there is a thermal lag in the cooling period, grid conductor temperature rises much faster than it decreases.

Assuming no reclosing operations and substituting the values for T_m and T_a in Eq. (8.5), we have

$$T_m = 1083 \quad T_a = 40 \quad \left(\frac{A}{I}\right)_{copper} = \sqrt{\frac{33 \cdot t}{\log\left[(T_m + 234) \div (T_a + 234)\right]}}$$

$t := \text{time}$

$$\log \frac{1317}{274} = 0.682 \qquad \sqrt{\frac{33}{0.682}} = 6.96$$

$$\left(\frac{A}{I}\right)_{copper} = 6.96 \times \sqrt{t} \qquad \text{circular mils/ampere} \tag{8.6}$$

Equations (8.5) and (8.6) are only applicable to copper conductors. Equation (8.7), introduced by B. Thapar, is a generalized formula applicable to any conductor material.

$$\frac{A}{I} = \left[\frac{4050.53 \, \rho_m \cdot \alpha \cdot t}{\gamma \cdot C \cdot \log[(1+\alpha \cdot T_m) \div (1+\alpha \cdot T_a)]} \right]^{0.5} \quad \text{circular mils/ampere}$$

(8.7)

where ρ_m = metal resistivity, in microhm-centimeters ($\mu\Omega \cdot$ cm)
α = metal temperature coefficient of resistance per degree Celsius
γ = metal specific weight, grams per cubic centimeter (g/cm³)
C = metal specific heat, calories per gram per degree Celsius (cal/g · °C)
T_m = maximum allowed temperature (°C)
T_a = conductor temperature before ground fault occurs (°C)

Table 8-2 provides the value of the data listed above for copper and steel.

Properties	Standard Annealed Copper	Typical Steel
ρ_m	1.724	17.4
α	0.00393	0.0042
γ	8.89	7.83
C	0.0928	0.118
T_m	1083	620*
T_a	40	40

*Indian practice for welded steel conductors.

TABLE 8-2 Copper and Steel Data

Substituting the values listed in Table 8-2 for copper in Eq. (8.7) we get:

$$\left(\frac{4050.53 \cdot 1.724 \cdot 0.00393}{8.89 \cdot 0.0928 \cdot \log\left(\frac{1+0.00393 \cdot 1083}{1+0.00393 \cdot 40}\right)} \right)^{0.5} = 7.11 \quad \left(\frac{A}{I}\right)_{copper} = 7.11 \cdot \sqrt{t}$$

(8.8)

Substituting the values listed in Table 8-2 for steel in Eq. (8.7) we get:

$$\left(\frac{4050.53 \cdot 17.4 \cdot 0.0042}{7.83 \cdot 0.118 \cdot \log\left(\frac{1+0.0042 \cdot 620}{1+0.0042 \cdot 40}\right)}\right)^{0.5} = 25.59 \quad \left(\frac{A}{I}\right)_{steel} = 25.59 \cdot \sqrt{t}$$

(8.9)

The circular mils to ampere ratio required to avoid melting, obtained using the generalized Eq. (8.7) for a cooper conductor, is only 2 percent larger than the result obtained using Eq. (8.6), which is the one recommended by the IEEE. Often the conductors that carry the highest current density during a substation ground fault are the lead conductors connecting an equipment enclosure or structural member to ground. Once each lead joins the grid, the short circuit current is split in many directions, including ground.

$$\frac{7.11}{6.96} \times 100 = 102.155$$

The cross-sectional area of steel conductors (maximum temperature 620°C) required to avoid fusing is 3.6 times larger than the cross-sectional area required for copper conductors (maximum temperature 1083°C). The international standard IEC 364 is based on a maximum conductor temperature of only 200°C for bare copper conductors.

8.6 Gradient Control

Let us start by summarizing the definitions of step and touch voltages and introducing the definitions of mesh and transferred voltages.

Step voltage is the potential difference between two points of the substation surface that are separated by 1 meter.

Tolerable step voltage is the maximum potential difference that a person can withstand when the body current path is from foot to foot. It is given by Eq. (7.5).

Touch voltage is the potential difference, during a fault to ground, between the place on the earth surface where a person may be standing and any grounded metal object than can be touched with either of the person's hands.

Tolerable touch voltage is the maximum potential difference that a person can withstand when the body current path is from either hand to both feet. It is given by Eq. (7.7).

Mesh voltage is the maximum potential difference, during a fault to ground, between the mesh conductor and a point on the earth surface within the mesh. For convenience, it is generally accepted that the point on the earth surface that maximizes the potential difference is the center of mesh. This assumption is rigorously correct only if the mat has a single mesh.

Transferred potential results when a fault to ground occurs inside the substation, and the short circuit current flowing through the ground mat impedance to earth produces a rise in the earth potential. Any metallic pipe buried in the substation surroundings will experience a rise in its potential and transfer this potential to distant places, such as building structures and remote perimeter fences. The transferring agent does not need to be buried in the ground; actually it could be any conducting object, such as shield wires, rails, telephone lines, or different types of communication equipment connected to the substation grounding system that become energized to the ground mat potential.

The ground fault could also occur in a tower of a high-voltage transmission line connected to the substation. In this case the transmission line shield wire transfers the potential rise to the substation ground mat. Something similar occurs if the ground fault happens in a power generating plant connected to the substation. In any of these cases the transferred potential could produce dangerous voltage gradients in remote places in the form of touch and step potentials.

In accordance with P. Laurent grid designed with average conductor spacing, diameter, and depth of burial, the typical range of values for the touch and step voltages are the following:

$$0.1\rho \times \frac{I}{L} \leq V_{step} \leq 0.15 \times \rho \times \frac{I}{L} \tag{8.10}$$

$$0.6 \times \rho \times \frac{I}{L} \leq V_{touch} \leq 0.8 \times \rho \times \frac{I}{L} \tag{8.11}$$

Formula (8.12) provides the mesh voltage; this simple equation assumes that the current flowing into earth per unit length is constant through the entire ground mat.

$$V_{mesh} = \rho \times \frac{I}{L} \quad \text{volt} \tag{8.12}$$

The required total length of grid conductors can be determined by equating the tolerable touch voltage, given by Eq. (7.7), to the approximated value of the mesh voltage given by Eq. (8.12). Symbolically,

$$\frac{116 + 0.17 \times \rho_s}{\sqrt{t}} = \rho \times \frac{I}{L} \tag{8.13}$$

$$L = \frac{\rho \times I \times \sqrt{t}}{116 + 0.17 \times \rho_s} \quad \text{meter} \tag{8.14}$$

Empirical equations for the mesh and step voltages, applicable to square ground mats formed with equal square meshes, are given below. The nonuniform leakage current distribution along the grid conductors is taken care of by introducing the k_i correction factor in both equations. In addition, Eq. (8.15) contains coefficient k_s and Eq. (8.16) contains coefficient k_m; both depend on the mat geometry, conductor's characteristic, and installation details.

$$V_{step} = k_s \times k_i \times \rho \times \frac{I}{L} \quad \text{volt} \tag{8.15}$$

$$V_{mesh} = k_m \times k_1 \times \rho \times \frac{I}{L} \quad \text{volt} \tag{8.16}$$

$$\frac{116 + 0.17 \times \rho_s}{\sqrt{t}} = k_m \times k_1 \times \rho \times \frac{I}{L} \tag{8.17}$$

$$L = \frac{k_m \times k_1 \times \rho \times I \times \sqrt{t}}{116 + 0.17 \times \rho_s} \quad \text{meter} \tag{8.18}$$

where I = total current flowing into earth from ground grid, ampere
L = total length of buried grid conductors, meter
ρ = average earth resistivity, ohm meter
ρ_s = earth surface resistivity, ohm meter
k_i = nonuniform current distribution correction factor
k_m = geometric factor applicable to mesh voltage computation
k_s = geometric factor applicable to step voltage computation

Equation (8.19) is an empirical formula that provides k_i, and Eq. (8.20) defines k_m. The initial theoretic concept was that if the grid was completely symmetric and the earth soil was perfectly homogeneous, then the leakage current flowing into the earth per unit length of grid conductor was constant through the entire grid. This implied that in case of a ground fault, the earth surface potential in the area covered by the grid is the same regardless of position. The only things that could affect the earth potential were the depth of burial and the number, size, and spacing of the grid conductors. It was later realized that this concept was not correct, that even with all these assumptions the leakage current increases toward the ends of each conductor and that the highest surface potential occurs on top of the corner meshes.

$$k_1 = 0.65 + 0.172 \times n \tag{8.19}$$

$$k_m = \frac{1}{2 \times \pi} \times \ln \frac{D^2}{16 \times h \times d} + \frac{1}{\pi} \times \ln \left(\prod_{i=3}^{n} \frac{2 \times i - 3}{2 \times i - 2} \right) \tag{8.20}$$

where D = grid conductor spacing, meters
d = grid conductor diameter, meters
h = depth of burial of grid conductors, meters
n = number of parallel grid conductors in one direction

Furthermore, Eq. (8.20) is only valid when the mesh voltage is computed for the center of the mesh. However, this location does not necessarily maximize the potential difference between the grid conductor and the earth surface. In case of a ground fault, the current flowing into earth per unit length of grid conductors is not constant. It increases from the conductor center toward the conductor ends. For a square grid with a symmetrical mesh pattern, the mesh voltage increases from the center of the mat toward the periphery. The numbers in Fig. 8-4 were experimentally determined by W. Koch, working with a scale model in a water tank. It shows the mesh voltages, in percent of conductor potentials, for a square grid with a symmetrical mesh pattern.

$$N = \frac{V_{mesh}}{E_{cond}} \times 100$$

$$\frac{N}{100} = \frac{k_m \times k_i \times \rho \times I/L}{R \times I} = \frac{k_m \times k_i \times \rho}{R \times L}$$

$$k_m \times k_i = \frac{N}{100} \times \frac{R}{\rho} \times L \qquad (8.21)$$

where N is the number shown inside the meshes in Fig. 8-4 and R is the ground grid resistance to earth, in ohms.

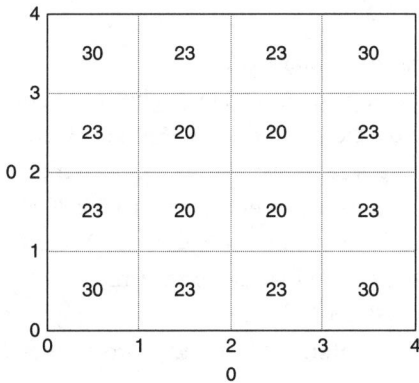

Figure 8-4 Mesh voltages in percent of conductor potentials.

If the product $k_m \times k_i$ is experimentally determined, then the value of k_j can be computed by first applying Eq. (8.20) and finding k_m and then using Eq. (8.22).

$$k_i = \frac{k_m \times k_i}{k_m} \qquad (8.22)$$

To compute the step voltage given by Eq. (8.15), it is necessary to apply Eq. (8.19) to determine k_i and Eq. (8.23) to determine k_s.

$$k_s = \frac{1}{\pi}\left[\frac{1}{2 \times h} + \frac{1}{D+h} + \sum_{i=3}^{n}\frac{1}{(i-1) \times D}\right] \qquad (8.23)$$

where D = grid conductor spacing, meters
d = grid conductor diameter, meters
h = depth of burial of grid conductor, meters
n = number of parallel grid conductors in one direction

Equation (8.23) was deduced assuming that the surface potential gradient remains constant at the maximum value and that this maximum gradient occurs for 1-meter long steps. Equation (8.23) is applicable to any ground mat where the conductor spacing is much larger than the depth of burial.

Step voltage computations do not require great accuracy because the touch voltage is the criterion used in the design of ground mats and the step voltage is only computed to check out the grid design, including the thickness of the crushed stone surface layer.

8.7 Example of Preliminary Grid Design

Data are as follows:

Mat configuration: square grid
Conductor spacing: same spacing in both directions
Area covered by the ground mat: $36 \times 36 = 1296$ m^2
Substation soil average resistivity: 300 ohm meter
Surface layer resistivity (wet crushed rock): 3000 ohm meter
Adjusted total fault current into grid: 1200 rms A
Fault clearing time: 100 milliseconds (ms)
Line voltage: 230 kV
Grid conductor material: stranded, bare copper cable with exothermic joints
Ground rods: copper clad steel, 1-in diameter, 6.1 meter long

Design Procedure

For this location the human's tolerable potentials are

$$E_{step} = \frac{116 + 0.7 \times \rho_s}{\sqrt{t}} = \frac{116 + 0.7 \times 3000}{\sqrt{0.1}} = 7008 \text{ volts}$$

$$E_{touch} = \frac{116 + 0.17 \times \rho_s}{\sqrt{t}} = \frac{116 + 0.17 \times 3000}{\sqrt{0.1}} = 1980 \text{ volts}$$

$$E_{let\text{-}go} = (1000 + 1.5 \times \rho_s) \times 0.009 = (1000 + 1.5 \times 3000) \times 0.009 = 49.5 \text{ volts}$$

The initial computation [see Eq. (8.14)] of the length of the grid conductors required is

$$L = \frac{\rho \times I \times \sqrt{t}}{116 + 0.17 \times \rho_s} = \frac{300 \times 1200 \times \sqrt{0.1}}{116 + 0.17 \times 3000} = 182 \text{ meter}$$

Practical experience shows that the usual range of conductor spacing and depths of burial are as follows:

Conductor spacing: 3 to 7.5 meter

Depth of burial: 0.5 to 1 meter

From the geometry of the mat and experience, the following initial conductor spacing, depth of burial, and number of rods are chosen.

$$D := 6 \text{ meter} \qquad h := 1 \text{ meter} \qquad n_r := 8$$

Ground rods do not improve appreciably the safety or the resistance to ground provided by the grid horizontal conductors; however, in case of prolonged drought, ground rods are very important. That is why it is better to use a few long rods than a bunch of short ones. See Fig. 8-5 for the preliminary grid layout.

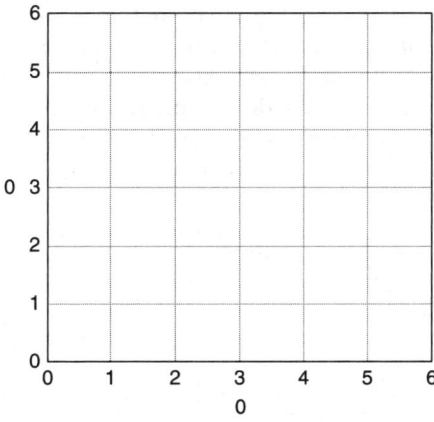

FIGURE 8-5 Initial grid design.

Chapter Eight

One rod in each corner and four more in between them, all connected to the perimeter conductors. Rod's coordinates: (0,0), (0,3), (0,6), (3,6), (6,6), (6,3), (6,0), and (3,0).

To calculate the conductor size, using Eq. (8.6), we have

$$\left(\frac{A}{I}\right) = 6.96 \times \sqrt{t} \quad \text{circular mils/ampere}$$

t = 4 second   Design thermal limit. Four seconds is a safe value because the protection will interrupt the fault in less than 100 milliseconds.

$6.96 \times \sqrt{4} = 13.92$ circular mils/ampere

I := 1200 ampere assuming that fault current is carried by one conductor

$A = 13.92 \times 1200 = 16{,}704$ circular mils

Because of mechanical ruggedness and the magnitude of the ground fault current leaking to earth, the only bare copper cable sizes considered are 1/0, 2/0, 3/0, and 4/0. Of these the most efficient with respect to resistance to ground and maximum mesh voltage is the 2/0 size.

$$2/0 \text{ AWG} = 133100 \text{ circular mils}$$
$$d = 0.414 \text{ inch (in)} = 0.0105 \text{ meter}$$

The time limit for the 2/0 copper cable is

$$6.96 \times \sqrt{t} = \frac{133{,}100}{1200} = 111 \quad \left(\frac{111}{6.96}\right)^2 = 254 \quad t = 254 \text{ seconds}$$

The maximum time that the 2/0 cable can withstand the ground fault current before melting is 254 seconds. Compared with 0.1 second which is the time it takes the fault to clear, this results in a safety margin of 253.9 seconds. The extra time provided by the 2/0 cable, together with the fact that in many cases the fault current is carried to ground by more than one conductor, provides protection against the breaker reclosing, operating mode, a very common case. From Eq. (8.18) we calculate again the total length of grid conductors required.

$$L = \frac{k_m \times k_1 \times \rho \times I \times \sqrt{t}}{116 + 0.17 \times \rho_s} \quad \text{meter} \qquad \text{See Eq. (8.18).}$$

$$k_i = 0.65 + 0.172 \times n \qquad \text{See Eq. (8.19).}$$

$$k_m = \frac{1}{2 \times \pi} \times \ln\frac{D^2}{16 \times h \times d} + \frac{1}{\pi} \times \ln\left(\prod_{i=3}^{n} \frac{2 \times i - 3}{2 \times i - 2}\right) \qquad \text{See Eq. (8.20).}$$

Ground Grid Preliminary Design

$n := 7 \quad \rho := 300 \quad \rho_s := 3000 \quad d := 0.0105 \quad h := 1 \quad t := 0.1 \quad D := 6$

$0.65 + 0.172 \times n = 1.854 \qquad k_i = 1.854$

$$\frac{1}{2 \times \pi} \times \ln \frac{D^2}{16 \times h \times d} + \frac{1}{\pi} \times \ln \left(\prod_{i=3}^{n} \frac{2 \times i - 3}{2 \times i - 2} \right) = 0.6009$$

$k_m = 0.6 \quad I := 1200$

$$\frac{0.6 \times 1.854 \times 300 \times 1200 \times \sqrt{0.1}}{116 + 0.17 \times 3000} = 202.3 \qquad L = 202.3 \text{ meter}$$

The required 202.3-meter length is 11 percent longer than the 182 meters initially computed. But the actual conductor length in the layout is $14 \times 36 = 504$ meters. Using Eq. (8.16), we obtain the mesh voltage.

$$V_{mesh} = k_m \times k_i \times \rho \times \frac{I}{L} \text{ Volt} \qquad \text{See Eq. (8.16).}$$

$$0.6 \times 1.854 \times 300 \times \frac{1200}{504} = 795 \qquad V_{mesh} = 795 \text{ volt}$$

$E_{touch} = 1980$ Volt   previously computed

The ground mat design shown in Fig. 8-5 complies with the conductors' length requirement of 202.3 meters. It actually provides 504 meters of conductors which is 2.5 times longer than the requirement without taking into consideration the length of the ground rods. Furthermore, the mesh voltage is only 795 Volts; that is 2.5 times smaller than the tolerable touch voltage, which is 1980 Volts. The values assumed in the initial attempt are very conservative because they provided much more than required, so let us try a 5×5 mesh grid design with the same perimeter dimensions, 36×36 meters, the same 2/0 copper cable, the same depth of burial, and the same number of rods.

Second Try

$\frac{36}{5} = 7.2$ spacing in both directions: $D := 7.2$ meter

$$L = \frac{k_m \times k_i \times \rho \times I \times \sqrt{t}}{116 + 0.17 \times \rho_s} \qquad \text{meter} \qquad \text{See Eq. (8.18).}$$

$k_i = 0.65 + 0.172 \times n$ See Eq. (8.19).

$$k_m = \frac{1}{2 \times \pi} \times \ln \frac{D^2}{16 \times h \times d} + \frac{1}{\pi} \times \ln \left(\prod_{i=3}^{n} \frac{2 \times i - 3}{2 \times i - 2} \right) \qquad \text{See Eq. (8.20).}$$

$n := 6 \quad D := 7.2$

Everything else has been defined before:

$\rho := 300 \quad \rho_s := 3000 \quad d := 0.0105 \quad h := 1 \quad t := 0.1$

$0.65 + 0.172 \times 6 = 1.682 \quad k_i = 1.682$

$\dfrac{1}{2 \times \pi} \times \ln \dfrac{7.2^2}{16 \times 1 \times 0.0105} + \dfrac{1}{\pi} \times \ln \left(\prod_{i=3}^{6} \dfrac{2 \times i - 3}{2 \times i - 2} \right) = 0.687 \quad k_m = 0.687$

From Eq. (8.18), we get the required conductor length.

$\dfrac{0.687 \times 1.682 \times 300 \times 1200 \times \sqrt{0.1}}{116 + 0.17 \times 3000} = 210 \quad L = 210 \text{ meter}$

Actual conductor length provided is $12 \times 36 = 432$ meters.

$V_{mesh} = k_m \times k_1 \times \rho \times \dfrac{I}{L} \quad \text{volt} \quad \text{See Eq. (8.16)}$

$0.687 \times 1.682 \times 300 \times \dfrac{1200}{432} = 963 \text{ volt} \quad V_{mesh} = 963 \text{ volt}$

$E_{touch} = 1980 \text{ volt} \quad$ previously computed tolerable touch voltage

The modified ground mat design with 5 × 5 meshes provides 432 meter of grid conductors which is twice the required length of 210 meters. And the tolerable touch voltage is twice the actual mesh voltage. Therefore, this is an acceptable design.

Computation of the Step Voltage Just outside the Corner Meshes

$V_{step} = k_s \times k_1 \times \rho \times \dfrac{I}{L} \qquad \text{See Eq. (8.15)}.$

$k_s = \dfrac{1}{\pi} \left[\dfrac{1}{2 \times h} + \dfrac{1}{D + h} \right] + \sum_{i=3}^{n} \dfrac{1}{(i-1) \times D} \qquad \text{See Eq. (8.23)}.$

$k_1 = 0.65 + 0.17 \times n \qquad \text{See Eq. (8.19)}.$

$\dfrac{1}{\pi} \left[\dfrac{1}{2 \times 1} + \dfrac{1}{7.2 + 1} \right] + \sum_{i=3}^{6} \dfrac{1}{(i-1) \times 7.2} = 0.25 \quad k_s = 0.25$

$0.65 + 0.17 \times 6 = 1.67 \quad$ average value for k_i

Ground Grid Preliminary Design

The nonuniform current distribution correction factor is assumed to be 1½ times larger than the average value of 1.67 previously computed, because the ground current density and voltage gradients are larger at the periphery and they are the largest at the corner meshes.

$k_1 = 2.5$ assumed value $1.5 \times 1.67 = 2.5$

$$V_{step} = 0.25 \times 2.5 \times 300 \times \frac{1200}{432}$$

$0.25 \times 2.5 \times 300 \times \frac{1200}{432} = 521$ maximum value of step voltage

The *tolerate* step voltage for a person walking on top of a layer of crushed rock with a resistivity not less than 3000 ohm meter was computed using Eq. (7.5). See design procedure.

$$\frac{116 + 0.7 \cdot 3000}{\sqrt{0.1}} = 7.008 \times 10^3 \text{ Volt} \qquad \frac{7.008 \times 10^3}{521} = 13.451$$

The tolerable step voltage is 13 times larger than the maximum value of the actual step voltage.

Return Ground Current Check

The corner mesh voltage produced by any return ground current must be smaller than the let-go potential. This current could be due to load unbalance. Someone may be touching the fence when the ground fault or load unbalance occurs.

Let-go potential = 49.5 Volts   previously calculated in design procedure

$$V_{mesh} = k_m \times k_1 \times \rho \times \frac{I_G}{L} \qquad \text{See Eq. (8.16)}$$

$\frac{0.687 \times 2.5 \times 300}{432} = 1.193 \qquad 1.193 \times I_G = 49.5$

$\frac{49.5}{1.193} = 41.5 \qquad I_G = 41.5 \text{ A}$

The substation protection ground relay's pickup current should be set for 40 A. This is a simplified computation; in a real case the calculations are more complicated.

Ground Mat Resistance to Remote Earth

$$R = \frac{\rho}{4 \times r} + \frac{\rho}{L} \quad \text{ohms} \quad \text{See Eq. (8.2).}$$

$$36^2 = \pi \times r^2 \quad \sqrt{\frac{36^2}{\pi}} = 20.3 \text{ meter} \quad r := 20.3 \quad \text{area: } 36 \times 36$$

$$\frac{300}{4 \times 20.3} + \frac{300}{432} = 4.389 \quad R = 4.389 \text{ ohms}$$

Maximum ground mat potential rise: $4.38 \times 1200 = 5256$ volts

CHAPTER 9
Principles of Ground Mat Design

9.1 Introduction

At any site selected for the construction of a substation, in general the earth is composed of several layers of soil; but in the design phase, the earth is usually represented by a one- or two-layer model. The two-layer model is a slightly better representation of the natural site, but it complicates the computations considerably, because the two-layer model generates an infinite number of images of the grid conductors that become weaker with increasing value of the z coordinate. This method was inspired by the well-known fact that an object placed between two mirrors generates an infinite number of images. But Huygens, Fresnel, and all these classical physicists were developing an exact science with concrete results that anyone can see and experience. This is very different from the artifice of assuming images formed by reflections of the electrical field at the boundary surface between the two soil layers with the purpose of eliminating the unpredictable effect of the ill-defined boundary.

During the construction phase, the natural earth structure is altered. The final result is a superficial layer of crushed stone: a layer of backfill material (original surface material plus material from the original first layer), the first layer of natural soil, and all other undisturbed soil layers existing at the site. Underground pipes, foundations, anchors, fences, drains, and culverts add complexity to the structure of the earth at the substation site. The system is very complex, and quite different from the single- or two-layer earth model. However, an exact representation would be extremely complex and totally impractical, when compared with available data. Ground grid designs assume that the soil layers are perfectly horizontal and of constant thickness and resistivity. Actually, there could be soil resistivity variations in the horizontal direction. Furthermore, the soil resistivity changes with temperature and experiences an abrupt increase when the soil temperature falls below 0°C. The soil resistivity also changes with the soil moisture content and experiences an abrupt increase when the moisture content of the soil decreases below 15 percent by weight.

The procedure for laying out the ground mat is as follows: The entire site is graded, trenches are open, grids' conductor intersections are cad-welded and lie inside the trenches; ground rods and equipment ground leads (two per each piece of equipment) are installed and connected to the grid. The trenches are backfilled, and a 4- or 6-inch-thick layer of crushed rock is placed over the entire site.

If the grid design calls for a 0.5-meter depth of burial, the installation contractor might decide not to open trenches, but instead to excavate the entire area to the required level and then layout the grid meshes with a layer of backfill material placed on top of the grid. Finally, a 4- or 6-inch-thick layer of crushed rock is placed over the entire site. This procedure pretty much invalidates the design calculations and must be avoided.

In normal mode of operation, the potential of the neutral of a power system is very close to zero; but when a fault to ground occurs, the potential of the neutral increases. This rise in potential is called a, *grounded metal potential rise* (GMPR). As a consequence of the fault and of the GMPR, a significant amount of current flows from the grid conductors into earth. The flow of current through the interface and soil impedances produce a voltage drop in the earth (the surface included), raising the potential of earth in the ground mat area with reference to remote earth. Of course, the mat area earth potential rise is not as large as the GMPR.

The frequency of the current flowing through earth is 60 Hz plus a decaying DC component due to the offset of the short circuit current wave form. Furthermore, capacitor banks could introduce very significant nonharmonics, low- and high-frequency components into the short circuit-to-ground current.

The depth of penetration of the ground fault current into earth is given by Eq. (9.1). The range of soil resistivity at substation sites is 50 to 1000 ohm-meters ($\Omega \cdot m$).

$f := 60$

$$\delta = \frac{1}{2 \cdot \pi} \cdot \sqrt{\frac{10^7 \cdot \rho}{f}} \qquad \text{meter} \qquad \text{current depth of penetration} \qquad (9.1)$$

$\rho := 50 \qquad \frac{1}{2 \cdot \pi} \cdot \sqrt{\frac{10^7 \cdot \rho}{f}} = 459.4407 \qquad \delta = 459 \text{ meters}$

$\rho := 1000 \qquad \frac{1}{2 \cdot \pi} \cdot \sqrt{\frac{10^7 \cdot \rho}{f}} = 2054.6815 \qquad \delta = 2055 \text{ meters}$

The skin depth or depth of penetration varies from 459 to 2055 meters. Because these numbers in most cases are more than twice the size of the longer dimensions of ground grid, which usually vary from

10 to 200 meters, it is acceptable to design the ground mat for direct current instead of alternating current. This means that the reactance of the grid conductors is neglected. This approach is acceptable because the available ground fault current is computed, most of the time, without taking into consideration the capacitors connected to the power system or the transmission line capacitance. Capacitors, especially the shunt-connected ones, introduce nonharmonic frequency oscillations in the short circuit wave shape.

9.2 Potential Created by a Point Current Source

First let us consider an infinitesimal segment of a buried metal conductor or a point current source with coordinates (x_s, y_s, z_s) from which the current I_s flows into homogeneous (constant resistivity) earth. The idea is to calculate the voltage produced by the point current source at any other point (x, y, z) inside the earth. The calculation must include the effect introduced by the earth surface which separates two regions of different resistivity. This mirrorlike boundary allows us to replace the real system depicted in Fig. 9-1

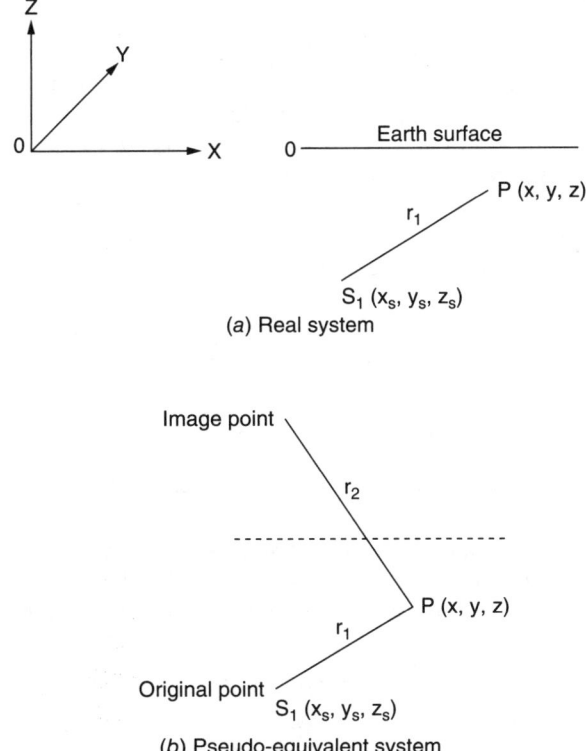

FIGURE 9-1 Method of images applied to a single point current source.

with a simpler, roughly equivalent one composed of two sources, the original and its image, both embedded in a homogeneous medium with no boundary surface. The voltage created at point (x,y,z) by each of the sources must comply with the Laplace equation. Let us consider now the potential function V(x,y,z) created by the original current source which is a differentiable scalar function of position. Laplace's homogeneous equation of this function is

$$\nabla \cdot (\nabla V) = 0 \quad \text{or} \quad \nabla^2 V = 0$$

where

$$\nabla^2 = \frac{\delta^2}{\delta x^2} + \frac{\delta^2}{\delta y^2} + \frac{\delta^2}{\delta z^2}$$

is the Laplacian operator in cartesian coordinates, V(x,y,z) is the potential function at point (x,y,z); therefore

$$\nabla^2 V = \frac{\delta^2 V}{\delta x^2} + \frac{\delta^2 V}{\delta y^2} + \frac{\delta^2 V}{\delta z^2} = 0 \qquad (9.2)$$

Equation (9.3) is a solution of Eq. (9.2) and is called the single-point-source solution. For a point current source of magnitude I_s in a homogeneous medium of resistivity ρ, the single-point-source solution is written as shown in Eq. (9.4), where Is is the rms value of the current, ρ is the resistivity of the homogeneous medium, and r is the distance from the point where the potential is calculated to the source. The potential created by the source is proportional to the current and inversely proportional to the distance.

$$V = \frac{1}{\sqrt{x^2 + y^2 + z^2}} = \frac{1}{r} \qquad (9.3)$$

$$V = \frac{\rho}{4 \cdot \pi} \cdot \frac{I_s}{r} \qquad (9.4)$$

Working with the equivalent system illustrated in Fig. 9-1 and superposing the potential created at point P by the original source and its image (both potentials are scalar quantities), we have

$$V = \frac{\rho \cdot I_s}{4 \cdot \pi} \left(\frac{1}{r_1} + \frac{1}{r_2} \right) \qquad (9.5)$$

The distances from point P(x,y,z) to the sources are

$S_1 = (x_s, y_s, z_s)$ source $r_1 = \sqrt{(x-x_S)^2 + (y-y_S)^2 + (z-z_S)^2}$

$S_2 = (x_s, y_s, -z_s)$ source image $r_2 = \sqrt{(x-x_S)^2 + (y-y_S)^2 + (z+z_S)^2}$

For the coordinate system selected, z and z_s are negative numbers because both the source and point P are below the earth surface. Substituting in Eq. (9.5), we obtain

$$V = \frac{\rho \cdot I_S}{4 \cdot \pi} \left[\frac{1}{\sqrt{(x-x_S)^2 + (y-y_S)^2 + (z-z_S)^2}} + \frac{1}{\sqrt{(x-x_S)^2 + (y-y_S)^2 + (z+z_S)^2}} \right]$$

(9.6)

Most ground grid designs are based on Eq. (9.6). It symbolically represents the expected result that the potentials induced are proportional to the source current and inversely proportional to the distances.

9.3 Potential at a Point inside Earth Created by Current Leaking to Earth from a Segment of a Grid Conductor

Figure 9-2 shows a horizontal square ground grid buried 1 meter below the earth surface. The coordinates of the random selected point and of segment-1 center are P(x,y,z) and (x_1, y_1, z_1). Segment 1 could be any of the grid segments. The current leaking out from the surface of segment 1 is assumed to be uniformly distributed (constant) along the segment, and the total current leaking out from segment 1 is designated as I_1. The diameter of the copper grid conductors usually ranges from 0.01 to 0.02 meter; these dimensions are small enough to treat any grid segment as a line current source. Ground grids are formed by square or rectangular meshes with conductors that are parallel or perpendicular to each other. The entire ground mat is supposed to be parallel to the ground surface. The coordinate system is selected in such a way that some conductors are oriented in the x direction, other's are oriented in the y direction and the ground rods are oriented in the z direction.

$\frac{I_1}{L_1}$ segment 1 current per unit length

164 Chapter Nine

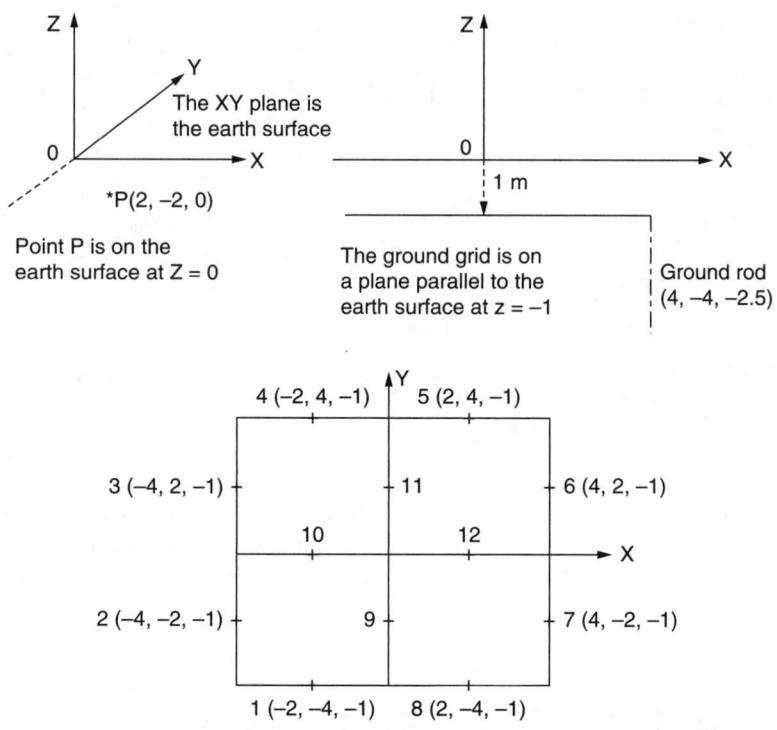

FIGURE 9-2 Square ground grid horizontally buried 1 meter below the earth surface.

$\dfrac{I_1 \cdot dx_S}{L_1}$ current leaking from an infinitesimal length of line source which in this case is oriented in x direction

(x_S, y_S, z_S) source coordinates

Applying Eq. (9.6), we obtain the potential produced at point $P(x, y, z)$ by an infinitesimal element of the source and its image.

x-directed segment:

$$dV_x = \frac{\rho \cdot I_1 \cdot dx_S}{4 \cdot \pi \cdot L_1} \left[\frac{1}{\sqrt{(x-x_S)^2 + (y-y_S)^2 + (z-z_S)^2}} + \frac{1}{\sqrt{(x-x_S)^2 + (y-y_S)^2 + (z+z_S)^2}} \right]$$

(9.7)

y-directed segment:

$$dV_y = \frac{\rho \cdot I_1 \cdot dy_S}{4 \cdot \pi \cdot L_1} \left[\frac{1}{\sqrt{(x-x_S)^2 + (y-y_S)^2 + (z-z_S)^2}} + \frac{1}{\sqrt{(x-x_S)^2 + (y-y_S)^2 + (z+z_S)^2}} \right]$$

(9.8)

z-directed segment:

$$dV_z = \frac{\rho \cdot I_1 \cdot dz_S}{4 \cdot \pi \cdot L_1} \left[\frac{1}{\sqrt{(x-x_S)^2 + (y-y_S)^2 + (z-z_S)^2}} + \frac{1}{\sqrt{(x-x_S)^2 + (y-y_S)^2 + (z+z_S)^2}} \right]$$

(9.9)

Let us solve the case of an *x-directed* source segment:

Coordinates of the line-source center: (x_1, y_1, z_1)
Source coordinates: (x_S, y_S, z_S) where $y_S = y_1, z_S = z_1$,
x_S variable along line source
Source segment length: L_1
Coordinates of the random selected point: $P(x, y, z)$

From Eq. (9.7) we obtain the potential at point P. See Eq. (9.10).

$$V_P = \frac{\rho \cdot I_1}{4 \cdot \pi \cdot L_1} \cdot \int_{x_1 - L_1/2}^{x_1 + L_1/2} \left[\frac{1}{\sqrt{(x-x_S)^2 + (y-y_S)^2 + (z-z_S)^2}} + \frac{1}{\sqrt{(x-x_S)^2 + (y-y_S)^2 + (z+z_S)^2}} \right] dx_S$$

(9.10)

For a specific case, the value of Eq. (9.10) is best found using numerical integration methods. The application of the formula, although repetitious, is simple and based on the data provided as follows:

For an x-directed segment: segment 8

$\rho := 200$ ohm-meter $I_1 := 1$ A $L_1 := 4$ meter
$x := 2$ $y := -2$ $z := 0$ point P (2, −2, 0) on earth surface
$x_1 := 2$ $y_1 := -4$ $z_1 := -1$ center of segment 8
$M_1(t, u) := \ln\left(t + \sqrt{t^2 + u^2}\right)$ Formula

$H_{xa} := \sqrt{(y - y_1)^2 + (z + z_1)^2}$ $H_{xa} = 2.2361$ $z + z_1 = -1$

$H_{xb} := \sqrt{(y - y_1)^2 + (z - z_1)^2}$ $H_{xb} = 2.2361$ $z - z_1 = 1$

$$V_P = \frac{\rho \cdot I_1}{4 \cdot \pi \cdot L_1} [M_{11}(x - x_1 + L_1, H_{xb}) - M_{12}(x - x_1 - L_1, H_{xb}) + M_{13}(x - x_1 + L_1, H_{xa}) - M_{14}(x - x_1 - L_1, H_{xa})]$$ (9.11)

Chapter Nine

Both M_{11} and M_{12} are derived from the first term of the definite integral Eq. (9.10), and M_{13} and M_{14} are derived from the second term of Eq. (9.10) and then evaluated by applying the given formula.

$u := H_{xb} = 2.2361$

$t := x - x_1 + L_1$ $t = 4$ $M_1(t, u) = 2.1497$ $M_{11} := 2.1497$

$t := x - x_1 - L_1$ $t = -4$ $M_1(t, u) = -0.5403$ $M_{12} := -0.5403$

$u := H_{xa} = 2.2361$

$t := x - x_1 + L_1$ $t = 4$ $M_1 = (t, u) = 2.1497$ $M_{13} := 2.1497$

$t := x - x_1 - L_1$ $t = -4$ $M_1(t, u) = -0.5403$ $M_{14} := -0.5403$

$M := M_{11} - M_{12} + M_{13} - M_{14}$ $M = 5.38$

$V_P := \dfrac{\rho \cdot I_1}{4 \cdot \pi \cdot L_1} \cdot M = 21.4063$ V_P is a scalar quantity, not a vector

For a y-directed segment: segment 7

$x_1 := 4$ $y_1 := -2$ $z_1 := -1$ $L_1 := 4$ meter   center of segment 7

$H_{ya} := \sqrt{(x - x_1)^2 + (z + z_1)^2}$ $H_{ya} = 2.2361$ $z + z_1 = -1$

$H_{yb} := \sqrt{(x - x_1)^2 + (z - z_1)^2}$ $H_{yb} = 2.2361$ $z - z_1 = 1$

$$V_P = \dfrac{\rho \cdot I_1}{4 \cdot \pi \cdot L_1} \big[M_{11}(y - y_1 + L_1, H_{yb}) - M_{12}(y - y_1 - L_1, H_{yb}) \\ + M_{13}(y - y_1 + L_1, H_{ya}) - M_{14}(y - y_1 - L_1, H_{ya}) \big] \quad (9.12)$$

$u := H_{yb} = 2.2361$

$t := y - y_1 + L_1$ $t = 4$ $M_1 = (t, u) = 2.1497$ $M_{11} := 2.1497$

$t := y - y_1 - L_1$ $t = -4$ $M_1(t, u) = -0.5403$ $M_{12} := -0.5403$

$u := H_{ya} = 2.2361$

$t := y - y_1 + L_1$ $t = 4$ $M_1(t, u) = 2.1497$ $M_{13} := 2.1497$

$t := y - y_1 - L_1$ $t = -4$ $M_1(t, u) = -0.5403$ $M_{14} := -0.5403$

$M := M_{11} - M_{12} + M_{13} - M_{14}$ $M = 5.38$

$V_P := \dfrac{\rho \cdot I_1}{4 \cdot \pi \cdot L_1} \cdot M = 21.4063$ same value as previous computation, because both segments are equal in length and are at same distance from point P

For a z-directed segment: ground rod

$x_1 := 4 \quad y_1 := -4 \quad z_1 := -2.5 \quad L_1 := 3 \quad$ center of ground rod

$H_{zb} := \sqrt{(x-x_1)^2 + (y-y_1)^2} \quad H_{zb} = 2.8284 \quad y - y_1 = 2$

$$V_p = \frac{\rho \cdot I_1}{4 \cdot \pi \cdot L_1} \cdot [M_{11}(z - z_1 + L_1, H_{zb}) - M_{12}(z - z_1 - L_1, H_{zb})$$
$$+ M_{13}(z + z_1 + L_1, H_{zb}) - M_{14}(z + z_1 - L_1, H_{zb})] \quad (9.13)$$

$u := H_{zb} = 2.8284$

$t := z - z_1 + L_1 \quad t = 5.5 \quad M_1(t, u) = 2.4583 \quad M_{11} := 2.4583$

$t := z - z_1 - L_1 \quad t = -0.5 \quad M_1(t, u) = 0.8639 \quad M_{12} := 0.8639$

$t := z + z_1 + L_1 \quad t = 0.5 \quad M_1(t, u) = 1.2156 \quad M_{13} := 1.2156$

$t := z + z_1 - L_1 \quad t = -5.5 \quad M_1(t, u) = -0.3788 \quad M_{14} := -0.3788$

$M := M_{11} - M_{12} + M_{13} - M_{14} \quad M = 3.1888$

$V_p := \frac{\rho \cdot I_1}{4 \cdot \pi \cdot L_1} \cdot M = 16.9171$

9.4 Mutual Resistance between Two Conductor Segments

The definition of the mutual resistance between two conductors, for instance, conductors 1 and 2, is based on Ohm's law, and it could be stated as the ratio of the voltage developed in conductor 2 per unit current in conductor 1. The application of this concept to ground grids is complicated by the fact that the segment currents are in fact the currents leaking to ground from the segments, and that all the segments are tied together and conducting current. The presentation is made with reference to Fig. 9-2. The problem is to find the potential created on segment 2 by each ampere of current leaking out from segment 1 and its image. Only segments 1 and 2 are considered; all other conductor segments are neglected because the mutual resistance is a one-to-one affair. In fact, there are 12 segments and one rod in Fig. 9-2. The effective mutual resistance in any one segment is composed of the mutual resistance due to the remaining 11 segments and their images and the rod and its image; some of these segments are parallel and others are perpendicular to the original segment. All this contributes to boring and long computations. When computing mutual resistances, it is an accepted procedure not to consider grounding rods at all—because the number of grounding rods leaking current to earth is much smaller than the number of segments leaking current, and because the distances separating rods and segments are larger than the distance between segments, with the exception of the segments on the periphery of the grid. However, we provided the equations and formulas necessary to carry out the computations including grounding rods. The task gets really complicated if the rods reach into a second layer of soil with different resistivity. The centers of segments 1 and 2 are at (x_1, y_1, z_1) and (x_2, y_2, z_2), and their lengths are L_1 and L_2, respectively. Assuming that segments 1 and 2 are x-directed and that I_1 is the total electric current leaking out from segment 1. This flow of current produces a potential in the centerline of segment 2 whose value per unit length or average is V_2. Applying Eq. (9.11) to segments 1 and 2 (for instance, segment 10 in Fig. 9-2), we obtain the potential produced at any point of the centerline of segment 2.

$$P_2 = \frac{\rho \cdot I_1}{4 \cdot \pi \cdot L_1}[M_1(x-x_1+L_1, H_{xb}) - M_1(x-x_1-L_1, H_{xb}) + M_1(x-x_1+L_1, H_{xa}) - M_1(x-x_1-L_1, H_{xa})]$$

where x is the x-coordinate of any point in the center line of segment 2, and x_1 is the x-coordinate of the center point of the source or segment 1, M_1 is defined as:

$$M_1(t, u) := \ln\left(t + \sqrt{t^2 + u^2}\right)$$

The average voltage induced in the centerline of segment 2 is

$$V_2 = \frac{1}{L_2} \cdot \int_{x_2-L_2/2}^{x_2+L_2/2} P_2 \, dx$$

$$V_2 = \frac{\rho \cdot I_1}{4 \cdot \pi \cdot L_1 \cdot L_2} \int_{x_2-L_2/2}^{x_2+L_2/2}[M_1(x-x_1+L_1, H_{xb}) - M_1(x-x_1-L_1, H_{xb})] \, dx$$

$$+ \frac{\rho \cdot I_1}{4 \cdot \pi \cdot L_1 \cdot L_2} \int_{x_2-L_2/2}^{x_2+L_2/2}[M_1(x-x_1+L_1, H_{xa}) - M_1(x-x_1-L_1, H_{xa})] \, dx$$

(9.14)

Equation (9.14) provides the average voltage developed in the centerline of segment 2. Its value could be found using the numerical integration procedure given below.

Two parallel, x-directed line segments: segments 1 and 10 of Fig. 9-2:

$\rho := 200 \qquad I_1 := 1$

$L_1 := 4 \quad L_2 := 4 \quad x_1 := -2 \quad y_1 := -4 \quad z_1 := -1 \quad x_2 := -2 \quad y_2 := 0$
$z_2 := -1$

Principles of Ground Mat Design

Segment 1: $u_{xb} := \sqrt{(y_2 - y_1)^2 + (z_2 - z_1)^2}$ $u_{xb} = 4$ $u_{xb}^2 = 16$

Image of 1: $u_{xa} := \sqrt{(y_2 - y_1)^2 + (z_2 + z_1)^2}$ $u_{xa} = 4.4721$ $u_{xa}^2 = 20$

$M_2(t, u) := t \cdot \ln\left(t + \sqrt{t^2 + u^2}\right) - \sqrt{t^2 + u^2}$ Formula

$M_{21}(x_2 - x_1 + L_1 + L_2, u_{xb}) - M_{22}(x_2 - x_1 + L_1 - L_2, u_{xb})$
$\quad - M_{23}(x_2 - x_1 - L_1 + L_2, u_{xb}) + M_{24}(x_2 - x_1 - L_1 - L_2, u_{xb})$
$\quad + M_{25}(x_2 - x_1 + L_1 + L_2, u_{xa}) - M_{26}(x_2 - x_1 + L_1 - L_2, u_{xa})$
$\quad - M_{27}(x_2 - x_1 - L_1 + L_2, u_{xa}) + M_{28}(x_2 - x_1 - L_1 - L_2, u_{xa})$ (9.15)

$M_2 = M_{21} - M_{22} - M_{23} + M_{24} + M_{25} - M_{26} - M_{27} + M_{28}$

$u := u_{xb} = 4$

$t := x_2 - x_1 + L_1 + L_2$ $t = 8$ $M_2(8, 4) = 13.6952$ $M_{21} := 13.695$

$t := x_2 - x_1 + L_1 - L_2$ $t = 0$ $M_2(0, 4) = -4$ $M_{22} := -4$

$t := x_2 - x_1 - L_1 + L_2$ $t = 0$ $M_2(0, 4) = -4$ $M_{23} := -4$

$t := x_2 - x_1 - L_1 - L_2$ $t = -8$ $M_2(-8, 4) = -8.4855$ $M_{24} := -8.486$

$u := u_{xa} = 4.4721$

$t := x_2 - x_1 + L_1 + L_2$ $t = 8$ $M_2(8, 4.472) = 13.5779$ $M_{25} := 13.578$

$t := x_2 - x_1 + L_1 - L_2$ $t = 0$ $M_2(0, 4.472) = -4.472$ $M_{26} := -4.472$

$t := x_2 - x_1 - L_1 + L_2$ $t = 0$ $M_2(0, 4.472) = -4.472$ $M_{27} := -4.472$

$t := x_2 - x_1 - L_1 - L_2$ $t = -8$ $M_2(-8, 4.472) = -10.3874$ $M_{28} := -10.387$

$M_2 := M_{21} - M_{22} - M_{23} + M_{24} + M_{25} - M_{26} - M_{27} + M_{28} = 25.344$

$r_{1.10} := \dfrac{\rho}{4 \cdot \pi \cdot L_1 \cdot L_2} \cdot M_2 = 25.2101$ mutual resistance between segments 1 and 10 is 25.21 ohms

For parallel y-directed segments, apply Eq. (9.16) and multiply the result as indicated in Eq. (9.17).

$M_3 = M_{31}(y_2 - y_1 + L_1 + L_2, u_{yb}) - M_{32}(y_2 - y_1 + L_1 - L_2, u_{yb})$

$\quad - M_{33}(y_2 - y_1 - L_1 + L_2, u_{yb}) + M_{34}(y_2 - y_1 - L_1 - L_2, u_{yb})$

$\quad + M_{35}(y_2 - y_1 + L_1 + L_2, u_{ya}) - M_{36}(y_2 - y_1 + L_1 - L_2, u_{ya})$

$\quad - M_{37}(y_2 - y_1 - L_1 + L_2, u_{ya}) + M_{38}(y_2 - y_1 - L_1 - L_2, u_{ya})$ (9.16)

$$u_{yb} = \sqrt{(x_2-x_1)^2 + (z_2-z_1)^2} \qquad u_{ya} = \sqrt{(x_2-x_1)^2 + (z_2+z_1)^2}$$

$$r_{y.y} = \frac{\rho}{4\cdot\pi\cdot L_1\cdot L_2}\cdot M_3 \tag{9.17}$$

$$M_3(t,u) := t\cdot\ln(t+\sqrt{t^2+u^2}) - \sqrt{t^2+u^2} \qquad \text{Formula} \tag{9.18}$$

For parallel z-directed segments, apply Eq. (9.19) and multiply the result as indicated in Eq. (9.20).

$$M_4 = M_{41}(z_2-z_1+L_1+L_2, u_{zb}) - M_{42}(z_2-z_1+L_1-L_2, u_{zb})$$

$$-M_{43}(z_2-z_1-L_1+L_2, u_{zb}) + M_{44}(z_2-z_1-L_1-L_2, u_{zb})$$

$$+M_{45}(z_2+z_1+L_1+L_2, u_{zb}) - M_{46}(z_2+z_1+L_1-L_2, u_{zb})$$

$$-M_{47}(z_2+z_1-L_1+L_2, u_{zb}) + M_{48}(z_2+z_1-L_1-L_2, u_{zb}) \tag{9.19}$$

$$u_{zb} = \sqrt{(x_2-x_1)^2 + (y_2-y_1)^2}$$

$$r_{z.z} = \frac{\rho}{4\cdot\pi\cdot L_1\cdot L_2}\cdot M_4 \tag{9.20}$$

$$M_4(t,u) := t\cdot\ln\left(t+\sqrt{t^2+u^2}\right) - \sqrt{t^2+u^2} \qquad \text{Formula} \tag{9.21}$$

For perpendicular segments (x-directed and y-directed), apply Eq. (9.22) and multiply the result as indicated in Eq. (9.23).

$$M_5 = M_{51} - M_{52} - M_{53} + M_{54} + M_{55} - M_{56} - M_{57} + M_{58} \tag{9.22}$$

$$M_{51}(x_2-x_1+L_1, y_2-y_1+L_2, z_2-z_1) - M_{52}(x_2-x_1+L_1, y_2-y_1-L_2, z_2-z_1)$$

$$-M_{53}(x_2-x_1-L_1, y_2-y_1+L_2, z_2-z_1) + M_{54}(x_2-x_1-L_1, y_2-y_1-L_2, z_2-z_1)$$

$$+M_{55}(x_2-x_1+L_1, y_2-y_1+L_2, z_2+z_1) - M_{56}(x_2-x_1+L_1, y_2-y_1-L_2, z_2+z_1)$$

$$-M_{57}(x_2-x_1-L_1, y_2-y_1+L_2, z_2+z_1) + M_{58}(x_2-x_1-L_1, y_2-y_1-L_2, z_2+z_1)$$

$$r_{x.y} = \frac{\rho}{4\cdot\pi\cdot L_1\cdot L_2}\cdot M_5 \tag{9.23}$$

Formula:

$$M_5(t,u,v) := -u + u\cdot\ln\left(t+\sqrt{t^2+u^2+v^2}\right) + t\cdot\ln\left(u+\sqrt{t^2+u^2+v^2}\right)$$

$$+ 2\cdot v\cdot\operatorname{atan}\frac{t+u+\sqrt{t^2+u^2+v^2}}{v} \tag{9.24}$$

Example 9-1
Compute the mutual resistance between segments 1 and 9. See Fig. 9-2.

$\rho := 200 \quad L_1 := 4 \quad L_2 := 4 \quad I_1 := 1$

$x_1 := -2 \quad y_1 := -4 \quad z_1 := -1 \quad x_2 := 0 \quad y_2 := -2 \quad z_2 = -1$

In the real world, the grid conductors' depths of burial are not constant. It is wiser to assume the following:

$z_2 = -0.96 \qquad z_2 - z_1 = 0.04 \qquad z_2 + z_1 = -1.96$

$v := z_2 - z_1 = 0.04$

$t := x_2 - x_1 + L_1 = 6$

$u := y_2 - y_1 + L_2 = 6 \qquad M_5(t, u, v) = 26.2032 \qquad M_{51} := 26.2032$

$t := x_2 - x_1 + L_1 = 6$

$u := y_2 - y_1 - L_2 = -2 \qquad M_5(t, u, v) = 5.8882 \qquad M_{52} := 5.8882$

$t := x_2 - x_1 - L_1 = -2$

$u := y_2 - y_1 + L_2 = 6 \qquad M_5(t, u, v) = -2.1118 \qquad M_{53} := -2.1118$

$t := x_2 - x_1 - L_1 = -2$

$u := y_2 - y_1 - L_2 \qquad M_5(t, u, v) = 2.6286 \qquad M_{54} := 2.6286$

$v := z_2 + z_1 = -1.96$

$t := x_2 - x_1 + L_1 = 6$

$u := y_2 - y_1 + L_2 = 6 \qquad M_5(t, u, v) = 32.0489 \qquad M_{55} := 32.0489$

$t := x_2 - x_1 + L_1 = 6$

$u := y_2 - y_1 - L_2 = -2 \qquad M_5(t, u, v) = 11.5555 \qquad M_{56} := 11.5555$

$t := x_2 - x_1 - L_1 = -2$

$u := y_2 - y_1 + L_2 = 6 \qquad M_5(t, u, v) = 3.5555 \qquad M_{57} := 3.5555$

$t := x_2 - x_1 - L_1 = -2$

$u := y_2 - y_1 - L_2 = -2 \qquad M_5(t, u, v) = -0.5506 \qquad M_{58} := -0.5506$

172 Chapter Nine

$$\frac{\rho}{4\cdot\pi\cdot L_1\cdot L_2}\cdot(M_{51}-M_{52}-M_{53}+M_{54}+M_{55}-M_{56}-M_{57}+M_{58})$$

$$=41.2238 \text{ ohms}$$

$r_{1.9} := 41.2238$ mutual resistance between segments 1 and 9

For perpendicular segments, x-directed and z-directed, apply Eq. (9.25) and multiply the result as indicated in Eq. (9.26).

$$M_6(t, u, v) := -u + u\cdot\ln\left(t+\sqrt{t^2+u^2+v^2}\right) + t\cdot\ln\left(u+\sqrt{t^2+u^2+v^2}\right)$$

$$+ 2\cdot v\,\text{atan}\left(\frac{t+u+\sqrt{t^2+u^2+v^2}}{v}\right) \tag{9.25}$$

$M_6 = M_{61} - M_{62} - M_{63} + M_{64} + M_{65} - M_{66} - M_{67} + M_{68}$

$M_{61}(x_2-x_1+L_1, z_2-z_1+L_2, y_2-y_1) - M_{62}(x_2-x_1+L_1, z_2-z_1-L_2, y_2-y_1)$

$-M_{63}(x_2-x_1-L_1, z_2-z_1+L_2, y_2-y_1) + M_{64}(x_2-x_1-L_1, z_2-z_1-L_2, y_2-y_1)$

$+M_{65}(x_2-x_1+L_1, z_2+z_1+L_2, y_2-y_1) - M_{66}(x_2-x_1+L_1, z_2+z_1-L_2, y_2-y_1)$

$-M_{67}(x_2-x_1-L_1, z_2+z_1+L_2, y_2-y_1) + M_{68}(x_2-x_1-L_1, z_2+z_1-L_2, y_2-y_1)$

$$r_{x.z} = \frac{\rho}{4\cdot\pi\cdot L_1\cdot L_2}\cdot M_6 \tag{9.26}$$

Example 9-2
Compute the mutual resistance between segment 12 (one) and the ground rod (two) shown in Fig. 9-2.

$\rho := 200$ $L_1 := 4$ $L_2 := 3$ $I_1 := 1$

$x_1 := 2$ $y_1 := 0$ $z_1 := -1$ $x_2 := 4$ $y_2 := -4$ $z_2 := -2.5$

$v := y_2 - y_1 = -4$

$t := x_2 - x_1 + L_1 = 6$

$u := z_2 - z_1 + L_2 = 1.5$ $M_6(t, u, v) = 25.9455$ $M_{61} := 25.9455$

$t := x_2 - x_1 + L_1 = 6$

$u := z_2 - z_1 - L_2 = -4.5$ $M_6(t, u, v) = 10.3064$ $M_{62} := 10.3064$

$t := x_2 - x_1 - L_1 = -2$

$u := z_2 - z_1 + L_2 = 1.5$ $M_6(t, u, v) = 2.8391$ $M_{63} := 2.8391$

$t := x_2 - x_1 - L_1 = -2$

$u := z_2 - z_1 - L_2 = -4.5$ $M_6(t, u, v) = -3.6453$ $M_{64} := -3.6453$

$t := x_2 - x_1 + L_1 = 6$

$u := z_2 + z_1 + L_2 = -0.5$ $M_6(t, u, v) = 20.7773$ $M_{65} := 20.7773$

$t := x_2 - x_1 + L_1 = 6$

$u := z_2 + z_1 - L_2 = -6.5$ $M_6(t, u, v) = 4.8801$ $M_{66} := 4.8801$

$t := x_2 - x_1 - L_1 = -2$

$u := z_2 + z_1 + L_2 = -0.5$ $M_6(t, u, v) = 0.9784$ $M_{67} := 0.9784$

$t := x_2 - x_1 - L_1 = -2$

$u := z_2 + z_1 - L_2 = -6.5$ $M_6(t, u, v) = -6.8954$ $M_{68} := -6.8954$

$$r_{x.z} := \frac{\rho}{4 \cdot \pi \cdot L_1 \cdot L_2} \cdot (M_{61} - M_{62} - M_{63} + M_{64} + M_{65} - M_{66} - M_{67} + M_{68})$$

$$= 22.7832$$

Mutual resistance between segment 12 and ground rod = 22.7832 ohms.

For perpendicular segments, y-directed and z-directed, apply Eq. (9.27) and multiply the result as indicated in Eq. (9.28).

Formula:

$$M_7(t, u, v) := -u + u \cdot \ln\left(t + \sqrt{t^2 + u^2 + v^2}\right)$$
$$+ t \cdot \ln\left(u + \sqrt{t^2 + u^2 + v^2}\right) + 2 \cdot v \cdot \mathrm{atan}\left(\frac{t + u + \sqrt{t^2 + u^2 + v^2}}{v}\right)$$

$$M_7 = M_{71} - M_{72} - M_{73} + M_{74} + M_{75} - M_{76} - M_{77} + M_{78} \quad (9.27)$$

where

$M_{71}(y_2 - y_1 + L_1, z_2 - z_1 + L_2, x_2 - x_1) - M_{72}(y_2 - y_1 + L_1, z_2 - z_1 - L_2, x_2 - x_1)$

$-M_{73}(y_2 - y_1 - L_1, z_2 - z_1 + L_2, x_2 - x_1) + M_{74}(y_2 - y_1 - L_1, z_2 - z_1 - L_2, x_2 - x_1)$

$+M_{75}(y_2 - y_1 + L_1, z_2 + z_1 + L_2, x_2 - x_1) - M_{76}(y_2 - y_1 + L_1, z_2 + z_1 - L_2, x_2 - x_1)$

$-M_{77}(y_2 - y_1 - L_1, z_2 + z_1 + L_2, x_2 - x_1) + M_{78}(y_2 - y_1 - L_1, z_2 + z_1 - L_2, x_2 - x_1)$

$$r_{y.z} = \frac{\rho}{4 \cdot \pi \cdot L_1 \cdot L_2} \cdot M_7 \quad (9.28)$$

9.5 Self-Resistance

The leakage self-resistance of a wire segment is the voltage produced in the segment by 1 ampere of its own leakage current or current leaking to earth. To simplify, the wire segment is considered to be a line-current source with uniform current and voltage. The voltage is the average voltage along the axis (line source) of the segment.

Segment length = L in meters
Segment radius = a in meters
Coordinates of the line-source center: (x_1, y_1, z_1) in meters

$$M_S(t, u) := t \cdot \ln\left(t + \sqrt{t^2 + u^2}\right) - \sqrt{t^2 + u^2} \qquad \text{Formula}$$

Self-resistance of x-directed or y-directed segments:

$$M_{S1}(L, a) + M_{S2}(-L, a) - 2a + M_{S3}\left(L, |z_1| \cdot \sqrt{2}\right) + M_{S4}\left(-L, |z_1| \cdot \sqrt{2}\right) - 2|z_1| \cdot \sqrt{2}$$

$$M_S = M_{S1} + M_{S2} - 2a + M_{S3} + M_{S4} - 2|z_1| \cdot \sqrt{2} \qquad (9.29)$$

$$r_S = \frac{\rho}{4 \cdot \pi \cdot L^2} \cdot M_S \qquad (9.30)$$

Example 9-3

L := 4	$z_1 := -1$	a := 0.00525	ρ := 200
t := L	t = 4		
u := a	u = 0.0053	$M_S(t, u) = 4.3178$	$M_{S1} := 4.3178$
t := −L	t := −4	$M_S(t, u) = 46.3018$	$M_{S2} := 46.3018$
t := L	t = 4		
$u := |z_1| \cdot \sqrt{2}$	u = 1.4142	$M_S(t, u) = 4.1946$	$M_{S3} := 4.1946$
t := −L	t = −4	$M_S(t, u) = 1.4221$	$M_{S4} := 1.4221$

$$M_S := \left(M_{S1} + M_{S2} - 2a + M_{S3} + M_{S4} - 2|z_1| \cdot \sqrt{2}\right) = 53.3974$$

$$r_S := \frac{\rho}{4 \cdot \pi \cdot L^2} \cdot M_S = 53.1153 \text{ ohms}$$

Principles of Ground Mat Design

Self-resistance of vertical (z-directed) segments:

$$M_S(t, u) := t \cdot \ln(t + \sqrt{t^2 + u^2}) - \sqrt{t^2 + u^2} \quad \text{Formula}$$

$$M_{S1}(L, a) + M_{S2}(-L, a) - 2a + M_{S3}(2 \cdot |z_1| + L, a) + M_{S4}(2 \cdot |z_1| - L, a) - 2 \cdot M_{S5}(2 \cdot |z_1|, a)$$

(9.31)

$$M_S = M_{S1} + M_{S2} - 2a + M_{S3} + M_{S4} - 2 \cdot M_{S5}$$

$$r_S = \frac{\rho}{4 \cdot \pi \cdot L^2} \cdot M_S \qquad (9.32)$$

Example 9-4

$\rho := 200 \qquad L := 3 \qquad z_1 := -2.5 \qquad a := 0.00952$

$u := a = 0.009525$

$t := L = 3 \qquad M_S(t, u) = 2.3753 \qquad M_{S.1} := 2.3753$

$t := -L = -3 \qquad M_S(t, u) = 30.2983 \qquad M_{S.2} := 30.2983$

$t := 2 \cdot |z_1| + L = 8 \qquad M_S(t, u) = 14.1807 \qquad M_{S.3} := 14.1807$

$t := 2 \cdot |z_1| - L = 2 \qquad M_S(t, u) = 0.7726 \qquad M_{S.4} := 0.7726$

$t := 2 \cdot |z_1| = 5 \qquad M_S(t, u) = 6.5129 \qquad M_{S.5} := 6.5129$

$M_S := M_{S.1} + M_{S.2} - 2 \cdot a + M_{S.3} + M_{S.4} - 2 \cdot M_{S.5} = 34.582$

$r_S := \frac{\rho}{4 \cdot \pi \cdot L^2} \cdot M_S = 61.1545 \text{ ohms}$

The self-resistance designation is misleading, because its value is independent of the segment material. It should be called *connection to earth resistance*.

When the leakage current distribution along the rod is uniform and the length of the rod is much larger than its diameter. Eq. (8.1) provides the resistance to earth of a single rod vertically buried in homogeneous soil.

$\rho := 200 \qquad L := 3 \qquad d := 0.01905$

$R_1 := \frac{\rho}{2 \times \pi \times L} \times \ln\left(\frac{4 \times L}{d}\right) \qquad \text{See Eq. (8.1)}$

$R_1 = 68.3899 \text{ ohms}$

In some applications, when the self-resistance is computed, the value resulting from this simple equation is close enough.

CHAPTER 10

Ground Mat Design with Nonuniform Current Distribution

10.1 Introduction

It is an experimental fact that in any ground grid the leakage current during a fault to ground increases from the center toward the periphery. And in any specific grid conductor the leakage current increases from the center toward the ends. In this chapter a method is presented to determine the grid leakage current distribution during a fault to ground and to calculate the surface potential at any given point.

10.2 Grid Current Distribution during a Fault to Ground

To study the leakage current distribution in a ground mat during a fault to ground, the grid is divided into segments. The leakage current along each of these segments is assumed constant, and the nonuniform grid leakage current distribution is taken care of by allowing it to change from one segment to the other. The pieces of conductors between nodes are arbitrarily selected as the segments mentioned above. These segments could be subdivided into smaller pieces, but this refinement complicates the computations and does not improve the results appreciably. The reader should keep in mind that the method presented below is only an approximation, and sometimes an approximation of an approximation.

Segments with equal leakage current are identified by geometric symmetry. This requires that the earth's resistivity be uniform over the entire extension of the grid. Figure 10-1 illustrates the procedure used to classify the segments. In this procedure and to facilitate the subsequent analysis, segments of the same type are numbered consecutively.

Chapter Ten

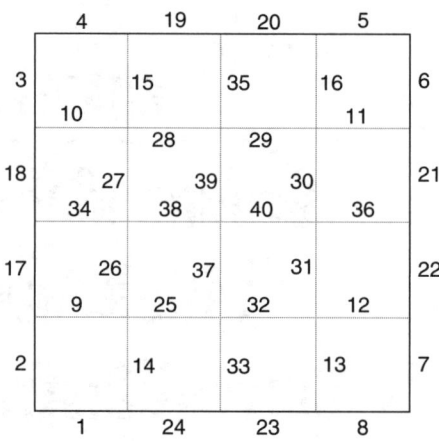

Figure 10-1 Six different currents.

As the list below indicates, there are six types of segments; those that belong to the same type have the same leakage current. Hence, there are only six different leakage currents, although the total number of segments is 40.

Type 1: end segments of an outside line	1 2 3 4 5 6 7 8
Type 2: end segments of a second-from-outside line	9 10 11 12 13 14 15 16
Type 3: second from the end segments of an outside line	17 18 19 20 21 22 23 24
Type 4: second from the end segments of a second-from-outside line	25 26 27 28 29 30 31 32
Type 5: end segments of a center line	33 34 35 36
Type 6: second from the end segments of a center line	37 38 39 40

In Fig. 10-1 all the segments have the same length because all the conductors are uniformly and symmetrically spaced, yielding six different types of segments. But if the east-west oriented conductors, e.g., were spaced symmetrically but not uniformly, then the number of different types of segments was 12 instead of 6.

The leakage currents in each type of segment are determined by solving simultaneously a set of independent linear equations. The following set of equations applies to the case shown in Fig. 10-1. The reader should be aware that all the conductor segments are tied together by earth and that the currents in the equations are leakage currents and not the conductor's current.

Ground Mat Design with Nonuniform Current Distribution

$$R_{11} \times I_1 + R_{12} \times I_2 + R_{13} \times I_3 + R_{14} \times I_4 + R_{15} \times I_5 + R_{16} \times I_6 = V_1$$
$$R_{21} \times I_1 + R_{22} \times I_2 + R_{23} \times I_3 + R_{24} \times I_4 + R_{25} \times I_5 + R_{26} \times I_6 = V_2$$
$$R_{31} \times I_1 + R_{32} \times I_2 + R_{33} \times I_3 + R_{34} \times I_4 + R_{35} \times I_5 + R_{36} + I_6 = V_3$$
$$R_{41} \times I_1 + R_{42} \times I_2 + R_{43} \times I_3 + R_{44} \times I_4 + R_{45} \times I_5 + R_{46} \times I_6 = V_4$$
$$R_{51} \times I_1 + R_{52} \times I_2 + R_{53} \times I_3 + R_{54} \times I_4 + R_{55} \times I_5 + R_{56} \times I_6 = V_5$$
$$R_{61} \times I_1 + R_{62} \times I_2 + R_{63} \times I_3 + R_{64} \times I_4 + R_{65} \times I_5 + R_{66} \times I_6 = V_6$$

(10.1)

where
R_{pp} = sum of self-resistance of a segment type P plus the mutual resistance between this segment and each of the remaining type P segments. For instance, let us consider segment 9 which is a type 2 segment. Then

$R_{22} = r_{9.9} + r_{9.10} + r_{9.11} + r_{9.12} + r_{9.13} + r_{9.14} + r_{9.15} + r_{9.16}$ The r's are the actual self-resistance and mutual resistance between segments.

R_{pq} = sum of mutual resistances between segment type p and all segments type q

For instance,

$R_{23} = r_{9.17} + r_{9.18} + r_{9.19} + r_{9.20} + r_{9.21} + r_{9.22} + r_{9.23} + r_{9.24}$ Segment 9 was selected to represent a type 2 segment.

V_p = voltage of any type segment with reference to remote earth
I_q = leakage current of any type of segment

The set of Eqs. (10.1) can be expressed as

$$\sum_{q=1}^{6}(R_{pq} \times I_q) = V_p \quad p := 1, 2..6$$

Generalizing the above expression, we obtain

$$\sum_{q=1}^{n}(R_{pq} \times I_q) = V_p \quad p := 1, 2.. n$$

In matrix notation,

$$\mathbf{RI = V}$$

This is a compact way of writing the following matrix equation:

$$\begin{pmatrix} R_{11} & R_{12} & R_{13} & R_{14} & R_{15} & R_{16} \\ R_{21} & R_{22} & R_{23} & R_{24} & R_{25} & R_{26} \\ R_{31} & R_{32} & R_{33} & R_{34} & R_{35} & R_{36} \\ R_{41} & R_{42} & R_{43} & R_{44} & R_{45} & R_{46} \\ R_{51} & R_{52} & R_{53} & R_{54} & R_{55} & R_{56} \\ R_{61} & R_{62} & R_{63} & R_{64} & R_{65} & R_{66} \end{pmatrix} \times \begin{pmatrix} I_1 \\ I_2 \\ I_3 \\ I_4 \\ I_5 \\ I_6 \end{pmatrix} = \begin{pmatrix} V_1 \\ V_2 \\ V_3 \\ V_4 \\ V_5 \\ V_6 \end{pmatrix} \quad (10.2)$$

The elements of matrix **V** are the segment voltages. However, it is acceptable to neglect the voltage drop along the grid conductors and to consider that all the segments are at the same potential, because at 60 hertz (Hz) the resistance and inductive reactance of the grid conductors are small compared to the grid resistance to earth. However, the inductance of the grid becomes important when a lightning discharge induces an electrical surge in the ground grid. Actually, the grid's inductance determines the grid impulse impedance. Furthermore, the voltage drops produced by leakage currents in the grid conductors must be taken into account only if the dimensions of the grid are almost equal or larger than the earth skin depth (current depth of penetration into earth) at 60 Hz. Then each segment voltage must be calculated as applied voltage minus voltage drop.

The formula to compute the leakage current depth of penetration in meters (m) is

$$\delta = \frac{1}{2 \times \pi} \times \sqrt{\frac{10^7 \times \rho}{f}} \quad \text{meters} \quad (10.3)$$

where ρ = earth resistivity in ohm-meters ($\Omega \cdot$ m) and f = current frequency in Hz.

The required number of equations to determine the leakage current in each type of segment is equal to the number of different types of segments. For a uniformly spaced square grid with n conductors in both directions, the number of different types of segments and the required number of equations are

If n is even: $\quad \dfrac{n^2}{4} \quad$ If n is odd: $\quad \dfrac{n^2 - 1}{4} \quad (10.4)$

10.3 Computations with Nonuniform Current Distribution in Small Square Grid

Figure 10-2 shows a 3 × 3 square ground grid with uniformly spaced conductors. It has three conductors in both directions with a total of 12 segments. The grid is buried in homogeneous soil with uniform resistivity.

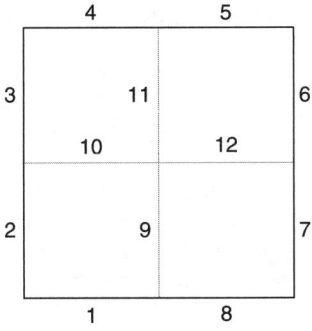

FIGURE 10-2 Uniformly spaced square grid.

Total number of segments = 12

$n := 3$ Odd number of conductors

$$\frac{n^2 - 1}{1} = 2$$ Number of different types of segments

x-directed   y-directed

1(–2, –4)   O   2(–4, –2)   O
4(–2, 4)   O   3(–4, 2)   O
5(2, 4)   O   6(4, 2)   O
8(2, –4)   O   7(4, –2)   O
10(–2, 0)   C   9(0, –2)   C
12(2, 0)   C   11(0, 2)   C

If the center point of a segment contains a 0, it is a center segment. If it contains a 4 it is an outside segment; O means outside segment and C means center segment.
    The design data are as follows:

$\rho := 200$ Ohm-meter, earth resistivity, homogeneous soil
$L := 4$ Meter, segment length
$a := 0.00526$ Meter, conductor radius; 2/0 bare copper
$Z := -1$ Meter, grid depth of burial
$V := 100$ Volt, segment potential to remote earth
$f := 60$ Hertz, voltage and current frequency

Segment Classification

The fact that the soil is uniform over the entire extension of the grid allows the classification of segments by geometric symmetry.

To facilitate the computation, segments that belong to the same type are numbered consecutively.

Type 1. End segments of an outside conductor: 1, 2, 3, 4, 5, 6, 7, 8
Type 2. End segments of a second from outside conductor: 9, 10, 11, 12

There are 12 segments, but only two different types. Therefore, only two different leakage currents must be computed. Those that belong to the same type have the same leakage current. The two leakage currents are determined by solving the following set of simultaneous linear equations:

$$R_{11} \times I_1 + R_{12} \times I_2 = V_1$$
$$R_{21} \times I_1 + R_{22} \times I_2 = V_2$$

or

$$\begin{pmatrix} R_{11} & R_{12} \\ R_{21} & R_{22} \end{pmatrix} \times \begin{pmatrix} I_1 \\ I_2 \end{pmatrix} = \begin{pmatrix} V_1 \\ V_2 \end{pmatrix}$$

Let us compare the largest dimension of the grid conductors to the current depth of penetration into earth.

$$\delta := \frac{1}{2 \times \pi} \times \sqrt{\frac{10^7 \times \rho}{f}} = 919 \text{ meter} \quad \text{current depth of penetration into earth}$$

Maximum length of one conductor in either direction = 8 meter $919 \gg 8$

Conductor voltage drop is negligible. Therefore, $V_1 = V_2 = V$.

The system of equations is reduced to

$$\begin{pmatrix} R_{11} & R_{12} \\ R_{21} & R_{22} \end{pmatrix} \times \begin{pmatrix} I_1 \\ I_2 \end{pmatrix} = \begin{pmatrix} V \\ V \end{pmatrix} \tag{10.5}$$

Determination of Matrix R Elements

The matrix **R** elements are a function of the leakage self-resistance and mutual resistance between segments. In this 3×3 grid, with 12 segments there are a total of $12^2 = 144$ self-resistances and mutual resistances. Fortunately, we don't need all of them to determine matrix **R** elements.

The matrix **R** elements as a function of the self-resistance and mutual resistances of the segments are

$R_{11} = r_{1.1} + r_{1.2} + r_{1.3} + r_{1.4} + r_{1.5} + r_{1.6} + r_{1.7} + r_{1.8}$ all segments type 1

$R_{22} = r_{9.9} + r_{9.10} + r_{9.11} + r_{9.12}$ all segments type 2

$R_{12} = r_{1.9} + r_{1.10} + r_{1.11} + r_{1.12}$ segments type 1 and type 2

$R_{21} = r_{9.1} + r_{9.2} + r_{9.3} + r_{9.4} + r_{9.5} + r_{9.6} + r_{9.7} + r_{9.8}$ segments type 2 and type 1

Ground Mat Design with Nonuniform Current Distribution

Self-resistance and mutual resistance are classified as follows:

Self-resistances: $r_{1.1}$ $r_{9.9}$

Parallel line segments: $r_{1.4}$ $r_{1.5}$ $r_{1.10}$ $r_{1.12}$ $r_{9.2}$ $r_{9.3}$ $r_{9.6}$ $r_{9.7}$
$r_{1.8}$ $r_{9.11}$

Perpendicular line segments: $r_{1.2}$ $r_{1.3}$ $r_{1.6}$ $r_{1.7}$ $r_{1.9}$ $r_{1.11}$
$r_{9.4}$ $r_{9.5}$ $r_{9.8}$ $r_{9.10}$ $r_{9.12}$

The mutual resistances between two line segments is reciprocal, for instance, $r_{19} = r_{9.1}$.

From the symmetry of Fig. 10-2 it is easy to agree with the following equalities. In this kind of task, humans are better than computers.

$r_{9.7} = r_{9.2} = r_{1.10} = 25.209$ $r_{9.8} = r_{9.1} = r_{1.9} = r_{1.2} = 41.223$

$r_{1.11} = r_{1.3} = 20.814$

$r_{9.6} = r_{9.3} = r_{1.12} = 21.722$ $r_{9.10} = r_{9.1} = r_{1.9} + r_{1.2} = 14.223$

$r_{1.1} = r_{9.9} = 53.112$

$r_{9.5} = r_{9.4} = r_{1.7} = 20.816$ $r_{9.12} = r_{9.10} = r_{9.1} = r_{1.9} = r_{1.2} = 41.223$

We will use Eq. (9.15) to compute all parallel x-directed line segments. See Fig. 10-2.

Segments 1 and 4, center of segment 1 (–2, –4, –1), center of segment 2 (–2, 4, –1)

$L_1 := 4$ $L_2 := 4$ $x_1 := -2$ $y_1 := -4$ $z_1 := -1$

$x_2 := -2$ $y_2 := 4$ $z_2 := -1$

To simplify the application of the formulas to any pair of segments, we always call them segments 1 and 2 regardless of their designation in the corresponding figures.

With the procedure and formulas presented in this chapter, it is possible to compute self and mutual resistances of any segment or pairs of segments. They are valid not only for the examples solved but also for the determination of self and mutual resistances in any grid in which the conductors are parallel or perpendicular to each other.

$$u_{xb} := \sqrt{(y_2 - y_1)^2 + (z_2 - z_1)^2} = 8$$

$$u_{xa} := \sqrt{(y_2 - y_1)^2 + (z_2 + z_1)^2} = 8.246$$

$$M_2(t, u) := t \times \ln\left(t + \sqrt{t^2 + u^2}\right) - \sqrt{t^2 + u^2} \qquad \text{Formula}$$

$$M_2 = M_{21} - M_{22} - M_{23} + M_{24} + M_{25} - M_{26} - M_{27} + M_{28}$$

Chapter Ten

$u := u_{xb} = 8$

$t := x_2 - x_1 + L_1 + L_2 = 8$ $M_2(t, u) = 12.373$ $M_{21} := 12.373$

$t := x_2 - x_1 + L_1 - L_2 = 0$ $M_2(t, u) = -8$ $M_{22} := -8$

$t := x_2 - x_1 - L_1 + L_2 = 0$ $M_2(t, u) = -8$ $M_{23} := -8$

$t := x_2 - x_1 - L_1 - L_2 = -8$ $M_2(t, u) = -20.898$ $M_{24} := -20.898$

$u := u_{xa} = 8.246$

$t := x_2 - x_1 + L_1 + L_2 = 8$ $M_2(t, u) = 12.27$ $M_{25} := 12.27$

$t := x_2 - x_1 + L_1 - L_2 = 0$ $M_2(t, u) = -8.246$ $M_{26} := -8.246$

$t := x_2 - x_1 - L_1 + L_2 = 0$ $M_2(t, u) = -8.246$ $M_{27} := -8.246$

$t := x_2 - x_1 - L_1 - L_2 = -8$ $M_2(t, u) = -21.486$ $M_{28} := -21.486$

$M_2 := M_{21} - M_{22} - M_{23} + M_{24} + M_{25} - M_{26} - M_{27} + M_{28} = 14.751$

$r_{1.4} := \dfrac{\rho}{4 \times \pi \times L_1 \times L_2} \times M_2 = 14.673 \text{ ohms}$ mutual resistance between segments 1 and 4

Segments 1 and 5, center of segment 1 (–2, –4, –1), center of segment 2 (2, 4, –1)

$x_2 := 2$ $y_2 := 4$ All the other coordinates remain the same as in segments 1 and 4.

$u_{xb} := \sqrt{(y_2 - y_1)^2 + (z_2 - z_1)^2} = 8$

$u_{xa} := \sqrt{(y_2 - y_1)^2 + (z_2 + z_1)^2} = 8.246$ $M_2(t, u) := t \times \ln\left(t + \sqrt{t^2 + u^2}\right) - \sqrt{t^2 + u^2}$

$u := u_{xb} = 8$

$t := x_2 - x_1 + L_1 + L_2 = 12$ $M_2(t, u) = 24.868$ $M_{21} := 24.868$

$t := x_2 - x_1 + L_1 - L_2 = 4$ $M_2(t, u) = 1.298$ $M_{22} := 1.298$

$t := x_2 - x_1 - L_1 + L_2 = 4$ $M_2(t, u) = 1.298$ $M_{23} := 1.298$

$t := x_2 - x_1 - L_1 - L_2 = -4$ $M_2(t, u) = -15.337$ $M_{24} := -15.337$

Ground Mat Design with Nonuniform Current Distribution

$u := u_{xa} = 8.246$

$t := x_2 - x_1 + L_1 + L_2 = 12$ $M_2(t, u) = 24.793$ $M_{25} := 24.793$

$t := x_2 - x_1 + L_1 - L_2 = 4$ $M_2(t, u) = 1.145$ $M_{26} := 1.145$

$t := x_2 - x_1 - L_1 + L_2 = 4$ $M_2(t, u) = 1.145$ $M_{27} := 1.145$

$t := x_2 - x_1 - L_1 - L_2 = -4$ $M_2(t, u) = -15.733$ $M_{28} := -15.733$

$M_2 := M_{21} - M_{22} - M_{23} + M_{24} + M_{25} - M_{26} - M_{27} + M_{28} = 13.705$

$r_{1.5} := \dfrac{\rho}{4 \times \pi \times L_1 \times L_2} \times M_2 = 13.633$ ohms   mutual resistance between segments 1 and 5

Segments 1 and 10, center of segment 1 (−2, −4, −1), center of segment 2 (−2, 0, −1)

$x_2 := -2$ $y_2 := 0$ All the other coordinates remain the same as in segments 1 and 4.

$u_{xb} := \sqrt{(y_2 - y_1)^2 + (z_2 - z_1)^2} = 4$

$u_{xa} := \sqrt{(y_2 - y_1)^2 + (z_2 + z_1)^2} = 4.472$ $M_2(t,u) := t \times \ln(t + \sqrt{t^2 + u^2}) - \sqrt{t^2 + u^2}$

$u := u_{xb} = 4$

$t := x_2 - x_1 + L_1 + L_2 = 8$ $M_2(t, u) = 13.695$ $M_{21} := 13.695$

$t := x_2 - x_1 + L_1 - L_2 = 0$ $M_2(t, u) = -4$ $M_{22} := -4$

$t := x_2 - x_1 - L_1 + L_2 = 0$ $M_2(t, u) = -4$ $M_{23} := -4$

$t := x_2 - x_1 - L_1 - L_2 = -8$ $M_2(t, u) = -8.486$ $M_{24} := -8.486$

$u := u_{xa} = 4.472$

$t := x_2 - x_1 + L_1 + L_2 = 8$ $M_2(t, u) = 13.578$ $M_{25} := 13.578$

$t := x_2 - x_1 + L_1 - L_2 = 0$ $M_2(t, u) = -4.472$ $M_{26} := -4.472$

$t := x_2 - x_1 - L_1 + L_2 = 0$ $M_2(t, u) = -4.472$ $M_{27} := -4.472$

$t := x_2 - x_1 - L_1 - L_2 = -8$ $M_2(t, u) = -10.388$ $M_{28} := -10.388$

$M_2 := M_{21} - M_{22} - M_{23} + M_{24} + M_{25} - M_{26} - M_{27} + M_{28} = 25.343$

$r_{1.10} := \dfrac{\rho}{4 \times \pi \times L_1 \times L_2} \times M_2 = 25.209$ ohms mutual resistance between segments 1 and 10

Segments 1 and 12, center of segment 1 (−2, −4, −1), center of segment 2 (2, 0, −1)

$x_2 := 2 \qquad y_2 := 0 \qquad$ All the other coordinates remain the same as in segments 1 and 4.

$u_{xb} := \sqrt{(y_2 - y_1)^2 + (z_2 - z_1)^2} = 4$

$u_{xa} := \sqrt{(y_2 - y_1)^2 + (z_2 + z_1)^2} = 4.472 \quad M_2(t, u) := t \times \ln(t + \sqrt{t^2 + u^2}) - \sqrt{t^2 + u^2}$

$u := u_{xb} = 4$

$t := x_2 - x_1 + L_1 + L_2 = 12 \qquad M_2(t, u) = 25.808 \qquad M_{21} := 25.808$

$t := x_2 - x_1 + L_1 - L_2 = 4 \qquad M_2(t, u) = 3.414 \qquad M_{22} := 3.414$

$t := x_2 - x_1 - L_1 + L_2 = 4 \qquad M_2(t, u) = 3.414 \qquad M_{23} := 3.414$

$t := x_2 - x_1 - L_1 - L_2 = -4 \qquad M_2(t, u) = -7.677 \qquad M_{24} := -7.677$

$u := u_{xa} = 4.472$

$t := x_2 - x_1 + L_1 + L_2 = 12 \qquad M_2(t, u) = 25.727 \qquad M_{25} := 25.727$

$t := x_2 - x_1 + L_1 - L_2 = 4 \qquad M_2(t, u) = 3.21 \qquad M_{26} := 3.21$

$t := x_2 - x_1 - L_1 + L_2 = 4 \qquad M_2(t, u) = 3.21 \qquad M_{27} := 3.21$

$t := x_2 - x_1 - L_1 - L_2 = -4 \qquad M_2(t, u) = -8.773 \qquad M_{28} := -8.773$

$M_2 := M_{21} - M_{22} - M_{23} + M_{24} + M_{25} - M_{26} - M_{27} + M_{28} = 21.837$

$\dfrac{\rho}{4 \times \pi \times L_1 \times L_2} \times M_2 = 21.722$ ohms mutual resistance between segments 1 and 12

Segments 1 and 8, center of segment 1 (−2, −4, −1), center of segment 2 (2, −4, −1)

$x_2 := 2 \qquad y_2 := -4 \qquad z_2 := -0.96 \qquad x_1 := -2 \qquad y_1 := -4 \qquad z_1 := -1$

$u_{xb} := \sqrt{(y_2 - y_1)^2 + (z_2 - z_1)^2} = 0.04$

Ground Mat Design with Nonuniform Current Distribution 187

$u_{xa} := \sqrt{(y_2 - y_1)^2 + (z_2 + z_1)^2} = 1.96$

$M_2(t, u) := t \times \ln(t + \sqrt{t^2 + u^2}) - \sqrt{t^2 + u^2}$

$u := u_{xb} = 0.04$

$t := x_2 - x_1 + L_1 + L_2 = 12$	$M_2(t, u) = 26.137$	$M_{21} := 26.137$
$t := x_2 - x_1 + L_1 - L_2 = 4$	$M_2(t, u) = 4.318$	$M_{22} := 4.318$
$t := x_2 - x_1 - L_1 + L_2 = 4$	$M_2(t, u) = 4.318$	$M_{23} := 4.318$
$t := x_2 - x_1 - L_1 - L_2 = -4$	$M_2(t, u) = 30.069$	$M_{24} := 30.069$

$u := u_{xa} = 1.96$

$t := x_2 - x_1 + L_1 + L_2 = 12$	$M_2(t, u) = 26.057$	$M_{25} := 26.057$
$t := x_2 - x_1 + L_1 - L_2 = 4$	$M_2(t, u) = 4.084$	$M_{26} := 4.084$
$t := x_2 - x_1 - L_1 + L_2 = 4$	$M_2(t, u) = 4.084$	$M_{27} := 4.084$
$t := x_2 - x_1 - L_1 - L_2 = -4$	$M_2(t, u) = -1.299$	$M_{28} := -1.299$

$M_2 := M_{21} - M_{22} - M_{23} + M_{24} + M_{25} - M_{26} - M_{27} + M_{28} = 64.16$

$r_{1.8} := \dfrac{\rho}{4 \times \pi \times L_1 \times L_2} \times M_2 = 63.821 \text{ ohms}$ mutual resistance between segments 1 and 8

Segments 9 and 11, center of segment 1 (0, –2, –1), center of segment 2 (0, 2, –1)

The mutual resistance between parallel y-directed segments will be computed using Eqs. (9.16), (9.17), and (9.18).

$x_2 := 0 \quad y_2 := 2 \quad z_2 := -0.96 \quad x_1 := 0 \quad y_1 := -2 \quad z_1 := -1$

$u_{yb} := \sqrt{(x_2 - x_1)^2 + (z_2 - z_1)^2} = 0.04$

$u_{ya} := \sqrt{(x_2 - x_1)^2 + (z_2 + z_1)^2} = 1.96 \quad M_3(t, u) := t \times \ln(t + \sqrt{t^2 + u^2}) - \sqrt{t^2 + u^2}$

$u := u_{yb} = 0.04$

$t := x_2 - x_1 + L_1 + L_2 = 8$	$M_3(t, u) = 14.181$	$M_{31} := 14.181$
$t := x_2 - x_1 + L_1 - L_2 = 0$	$M_3(t, u) = -0.04$	$M_{32} := -0.04$
$t := x_2 - x_1 - L_1 + L_2 = 0$	$M_3(t, u) = -0.04$	$M_{33} := -0.04$
$t := x_2 - x_1 - L_1 - L_2 = -8$	$M_3(t, u) = 65.683$	$M_{34} := 65.683$

$u := u_{ya} = 1.96$

$t := x_2 - x_1 + L_1 + L_2 = 8$ $\qquad M_3(t, u) = 14.062 \qquad M_{35} := 14.062$

$t := x_2 - x_1 + L_1 - L_2 = 0$ $\qquad M_3(t, u) = -1.96 \qquad M_{36} := -1.96$

$t := x_2 - x_1 - L_1 + L_2 = 0$ $\qquad M_3(t, u) = -1.96 \qquad M_{37} := -1.96$

$t := x_2 - x_1 - L_1 - L_2 = -8$ $\qquad M_3(t, u) = 3.294 \qquad M_{38} := 3.294$

$M_3 := M_{31} - M_{32} - M_{33} + M_{34} + M_{35} - M_{36} - M_{37} + M_{38} = 101.22$

$r_{9.11} := \dfrac{\rho}{4 \times \pi \times L_1 \times L_2} \times M_3 = 100.685$ ohms $\qquad$ mutual resistance between segments 9 and 11

Segments 1 and 2, center of segment 1 (−2, −4, −1), center of segment 2 (−4, −2, −0.96)

For perpendicular segments (x-directed and y-directed) apply Eqs. (9.22) and (9.24) and multiply the results as indicated in Eq. (9.23).

$x_2 := -4 \qquad y_2 := -2 \qquad z_2 := -0.96 \qquad L_2 := 4$

$x_1 := -2 \qquad y_1 := -4 \qquad z_1 := -1 \qquad L_1 := 4$

$M_5(t, u, v) := -u + u \times \ln\left(t + \sqrt{t^2 + u^2 + v^2}\right) + t \times \ln\left(u + \sqrt{t^2 + u^2 + v^2}\right)$

$\qquad\qquad + 2 \times v \times \operatorname{atan}\left(\dfrac{t + u + \sqrt{t^2 + u^2 + v^2}}{v}\right)$

$v := z_2 - z_1 = 0.04 \qquad$ division by zero not allowed

$t := x_2 - x_1 + L_1 = 2$

$u := y_2 - y_1 + L_2 = 6 \qquad M_5(t, u, v) = 11.864 \qquad M_{51} := 11.864$

$t := x_2 - x_1 + L_1 = 2$

$u := y_2 - y_1 - L_2 = -2 \qquad M_5(t, u, v) = -1.4 \qquad M_{52} := -1.4$

$t := x_2 - x_1 - L_1 = -6$

$u := y_2 - y_1 + L_2 = 6 \qquad M_5(t, u, v) = -16.451 \qquad M_{53} := -16.451$

$t := x_2 - x_1 - L_1 = -6$

$u := y_2 - y_1 - L_2 = -2 \qquad M_5(t, u, v) = -4.66 \qquad M_{54} := -4.66$

Ground Mat Design with Nonuniform Current Distribution

$v := z_2 + z_1 = -1.96$

$t := x_2 - x_1 + L_1 = 2$

$u := y_2 - y_1 + L_2 = 6$ $M_5(t, u, v) = 17.631$ $M_{55} := 17.631$

$t := x_2 - x_1 + L_1 = 2$

$u := y_2 - y_1 - L_2 = -2$ $M_5(t, u, v) = 3.471$ $M_{56} := 3.471$

$t := x_2 - x_1 - L_1 = -6$

$u := y_2 - y_1 + L_2 = 6$ $M_5(t, u, v) = -10.862$ $M_{57} := -10.862$

$t := x_2 - x_1 - L_1 = -6$

$u := y_2 - y_1 - L_2 = -2$ $M_5(t,u,v) = -8.635$ $M_{58} := -8.635$

$M_5 := M_{51} - M_{52} - M_{53} + M_{54} + M_{55} - M_{56} - M_{57} + M_{58} = 41.442$

$r_{1,2} := \dfrac{\rho}{4 \times \pi \times L_1 \times L_2} \times M_5 = 41.223$ ohms   mutual resistance between segments 1 and 2

Segments 1 and 3, center of segment 1 (−2, −4, −1), center of segment 3 (−4, 2, −0.96)

Apply Eqs. (9.22), (9.23), and (9.24).

$x_2 := -4$ $y_2 := 2$ $z_2 := -0.96$

$x_1 := -2$ $y_1 := -4$ $z_1 := -1$

$M_5(t, u, v) := -u + u \times \ln\left(t + \sqrt{t^2 + u^2 + v^2}\right) + t \times \ln\left(u + \sqrt{t^2 + u^2 + v^2}\right)$

$\qquad + 2 \times v \times \mathrm{atan}\left(\dfrac{t + u + \sqrt{t^2 + u^2 + v^2}}{v}\right)$

$v := z_2 - z_1 = 0.04$ division by zero not allowed

$t := x_2 - x_1 + L_1 = 2$

$u := y_2 - y_1 + L_2 = 10$ $M_5(t, u, v) = 21.15$ $M_{51} := 21.15$

$t := x_2 - x_1 + L_1 = 2$

$u := y_2 - y_1 - L_2 = 2$ $M_5(t, u, v) = 4.424$ $M_{52} := 4.424$

$t := x_2 - x_1 - L_1 = -6$

$u := y_2 - y_1 + L_2 = 10$ $M_5(t, u, v) = -10.99$ $M_{53} := -10.99$

$t := x_2 - x_1 - L_1 = -6$

$u := y_2 - y_1 - L_2 = 2$ $M_5(t,u,v) = -16.841$ $M_{54} := -16.841$

$v := z_2 + z_1 = -1.96$

$t := x_2 - x_1 + L_1 = 2$

$u := y_2 - y_1 + L_2 = 10$ $M_5(t, u, v) = 27.009$ $M_{55} := 27.009$

$t := x_2 - x_1 + L_1 = 2$

$u := y_2 - y_1 - L_2 = 2$ $M_5(t, u, v) 9.924$ $M_{56} := 9.924$

$t := x_2 - x_1 - L_1 = -6$

$u := y_2 - y_1 + L_2 = 10$ $M_5(t, u, v) = -5.202$ $M_{57} := 5.202$

$t := x_2 - x_1 - L_1 = -6$

$u := y_2 - y_1 - L_2 = 2$ $M_5(t,u,v) = -12.237$ $M_{58} := -12.237$

$M_5 := M_{51} - M_{52} - M_{53} + M_{54} + M_{55} - M_{56} - M_{57} + M_{58} = 20.925$

$r_{1.3} := \dfrac{\rho}{4 \times \pi \times L_1 \times L_2} \times M_5 = 20.814$ ohms   mutual resistance between segments 1 and 3

Segments 1 and 6, center of segment 1 (−2, −4, −1), center of segment 6 (4, 2, −1)

Apply Eqs. (9.22), (9.23), and (9.24).

$x_2 := 4$ $y_2 := 2$ $z_2 := -0.96$

$x_1 := -2$ $y_1 := -4$ $z_1 := -1$

$M_5(t, u, v) := -u + u \times \ln\left(t + \sqrt{t^2 + u^2 + v^2}\right) + t \times \ln\left(u + \sqrt{t^2 + u^2 + v^2}\right)$
$+ 2 \times v \times \operatorname{atan}\left(\dfrac{t + u + \sqrt{t^2 + u^2 + v^2}}{v}\right)$

$v := z_2 - z_1 = 0.04$ division by zero not allowed

$t := x_2 - x_1 + L_1 = 10$

$u := y_2 - y_1 + L_2 = 10$ $M_5(t, u, v) = 53.805$ $M_{51} := 53.805$

$t := x_2 - x_1 + L_1 = 10$

$u := y_2 - y_1 - L_2 = 2$ $M_5(t, u, v) = 29.15$ $M_{52} := 29.15$

$t := x_2 - x_1 - L_1 = 2$

$u := y_2 - y_1 + L_2 = 10$ $M_5(t, u, v) = 21.15$ $M_{53} := 21.15$

Ground Mat Design with Nonuniform Current Distribution

$t := x_2 - x_1 - L_1 = 2$

$u := y_2 - y_1 - L_2 = 2$ $\quad M_5(t, u, v) = 4.424 \quad\quad M_{54} := 4.424$

$v := z_2 + z_1 = -1.96$

$t := x_2 - x_1 + L_1 = 10$

$u := y_2 - y_1 + L_2 = 10$ $\quad M_5(t, u, v) = 59.724 \quad\quad M_{55} := 59.724$

$t := x_2 - x_1 + L_1 = 10$

$u := y_2 - y_1 - L_2 = 2$ $\quad M_5(t, u, v) = 35.009 \quad\quad M_{56} := 35.009$

$t := x_2 - x_1 - L_1 = 2$

$u := y_2 - y_1 + L_2 = 10$ $\quad M_5(t, u, v) = 27.009 \quad\quad M_{57} := 27.009$

$t := x_2 - x_1 - L_1 = 2$

$u := y_2 - y_1 - L_2 = 2$ $\quad M_5(t, u, v) = 9.924 \quad\quad M_{58} := 9.924$

$M_5 := M_{51} - M_{52} - M_{53} + M_{54} + M_{55} - M_{56} - M_{57} + M_{58} = 15.559$

$r_{1.6} := \dfrac{\rho}{4 \times \pi \times L_1 \times L_2} \times M_5 = 15.477$ ohms   mutual resistance between setments 1 and 6

Segments 1 and 7, center of segment 1 (−2, −4, −1), center of segment 7 (4, −2, −0.96).
   Apply Eqs. (9.22), (9.23), and (9.24).

$\quad\quad x_2 := 4 \quad\quad y_2 := -2 \quad\quad z_2 := -0.96$

$\quad\quad x_1 := -2 \quad\quad y_1 := -4 \quad\quad z_1 := -1$

$M_5(t, u, v) := -u + u \times \ln\left(t + \sqrt{t^2 + u^2 + v^2}\right) + t \times \ln\left(u + \sqrt{t^2 + u^2 + v^2}\right)$

$\quad\quad + 2 \times v \times \operatorname{atan}\left(\dfrac{t + u + \sqrt{t^2 + u^2 + v^2}}{v}\right)$

$v := z_2 - z_1 = 0.04$ division by zero not allowed

$t := x_2 - x_1 + L_1 = 10$

$u := y_2 - y_1 + L_2 = 6$ $\quad M_5(t, u, v) = 41.293 \quad\quad M_{51} := 41.293$

$t := x_2 - x_1 + L_1 = 10$

$u := y_2 - y_1 - L_2 = -2$ $\quad M_5(t, u, v) = 17.153 \quad\quad M_{52} := 17.153$

$t := x_2 - x_1 - L_1 = 2$

$u := y_2 - y_1 + L_2 = 6$ $\quad M_5(t, u, v) = 11.864 \quad M_{53} := 11.864$

$t := x_2 - x_1 - L_1 = 2$

$u := y_2 - y_1 - L_2 = -2$ $\quad M_5(t, u, v) = -1.4 \quad M_{54} := -1.4$

$v := z_2 + z_1 = -1.96$

$t := x_2 - x_1 + L_1 = 10$

$u := y_2 - y_1 + L_2 = 6$ $\quad M_5(t, u, v) = 47.187 \quad M_{55} := 47.187$

$t := x_2 - x_1 + L_1 = 10$

$u := y_2 - y_1 - L_2 = -2$ $\quad M_5(t, u, v) = 22.976 \quad M_{56} := 22.976$

$t := x_2 - x_1 - L_1 = 2$

$u := y_2 - y_1 + L_2 = 6$ $\quad M_5(t, u, v) = 17.631 \quad M_{57} := 17.631$

$t := x_2 - x_1 - L_1 = 2$

$u := y_2 - y_1 - L_2 = -2$ $\quad M_5(t, u, v) = 3.471 \quad M_{58} := 3.471$

$M_5 := M_{51} - M_{52} - M_{53} + M_{54} + M_{55} - M_{56} - M_{57} + M_{58} = 20.927$

$r_{1,7} := \dfrac{\rho}{4 \times \pi \times L_1 \times L_2} \times M_5 = 20.816$ ohms   mutual resistance between segments 1 and 7

Segments 4 and 9, center of segment 1 (−2, 4, −1), center of segment 2 (0, −2, −0.96)

$r_{4,9} = r_{1,7} = 20.816$ We know this from the symmetry of Fig. 10-2. We will verify this below.

$x_2 := 0 \qquad y_2 := -2 \qquad z_2 := -0.96 \qquad$ segment 9

$x_1 := -2 \qquad y_1 := 4 \qquad z_1 - 1 \qquad$ segment 4

$M_5(t, u, v) := -u + u \times \ln(t + \sqrt{t^2 + u^2 + v^2}) + t \times \ln(u + \sqrt{t^2 + u^2 + v^2})$

$\qquad + 2 \times v \times \operatorname{atan}\left(\dfrac{t + u + \sqrt{t^2 + u^2 + v^2}}{v}\right) \quad$ See Eq. (9.24)

$v := z_2 - z_1 = 0.04 \qquad$ division by zero not allowed

$t := x_2 - x_1 + L_1 = 6$

$u := y_2 - y_1 + L_2 = -2$
$t := x_2 - x_1 + L_1 = 6$ $\quad M_5(t, u, v) = 5.888 \quad M_{51} := 5.888$

$u := y_2 - y_1 + L_2 = -10 \quad M_5(t, u, v) = -15.541 \quad M_{52} := -15.541$

$t := x_2 - x_1 - L_1 = -2$

$u := y_2 - y_1 + L_2 = -2$ $M_5(t, u, v) = 2.629$ $M_{53} := 2.629$

$t := x_2 - x_1 - L_1 = -2$

$u := y_2 - y_1 - L_2 = -10$ $M_5(t, u, v) = -7.925$ $M_{54} := -7.925$

$v := z_2 + z_1 = -1.96$

$t := x_2 - x_1 + L_1 = 6$

$u := y_2 - y_1 + L_2 = -2$ $M_5(t, u, v) = 11.555$ $M_{55} := 11.555$

$t := x_2 - x_1 + L_1 = 6$

$u := y_2 - y_1 - L_2 = -10$ $M_5(t, u, v) = -10$ $M_{56} := -10$

$t := x_2 - x_1 - L_1 = -2$

$u := y_2 - y_1 + L_2 = -2$ $M_5(t, u, v) = -0.551$ $M_{57} := -0.551$

$t := x_2 - x_1 - L_1 = -2$

$u := y_2 - y_1 - L_2 = -10$ $M_5(t, u, v) = -12.055$ $M_{58} := -12.055$

$M_5 := M_{51} - M_{52} - M_{53} + M_{54} + M_{55} - M_{56} - M_{57} + M_{58} = 20.926$

$r_{4.9} := \dfrac{\rho}{4 \times \pi \times L_1 \times L_2} \times M_5 = 20.8155$ ohms   mutual resistance between segments 4 and 9

$r_{4.9} = r_{1.7} = 20.816$ The equality of these two mutual resistances was assumed first because of the symmetrical location of these two pairs of segments. This has now been verified by computations.

Segments 9 and 10, center of segment 1 (0, −2, −1), center of segment 2 (−2, 0, −0.96)

By symmetry, the mutual resistance of this pair of segments should be equal to the mutual resistance between segments 1 and 2 that has already been computed as 41.223 ohms. Again, we will verify this assumption.

$x_2 := -2$ $y_2 := 0$ $z_2 := -0.96$ segment 10

$x_1 := 0$ $y_1 := -2$ $z_1 := -1$ segment 9

$M_5(t, u, v) := -u + u \times \ln\left(t + \sqrt{t^2 + u^2 + v^2}\right) + t \times \ln\left(u + \sqrt{t^2 + u^2 + v^2}\right)$

$\qquad + 2 \times v \times \mathrm{atan}\left(\dfrac{t + u\sqrt{t^2 + u^2 + v^2}}{v}\right)$ See Eq. (9.24)

$v := z_2 - z_1 = 0.04$ division by zero not allowed

$t := x_2 - x_1 + L_1 = 2$

$u := y_2 - y_1 + L_2 = 6$ $M_5(t, u, v) = 11.864$ $M_{51} := 11.864$

$t := x_2 - x_1 + L_1 = 2$

$u := y_2 - y_1 - L_2 = -2$ $M_5(t, u, v) = -1.4$ $M_{52} := -1.4$

$t := x_2 - x_1 - L_1 = -6$

$u := y_2 - y_1 + L_2 = 6$ $M_5(t, u, v) = -16.451$ $M_{53} := -16.451$

$t := x_2 - x_1 - L_1 = -6$

$u := y_2 - y_1 - L_2 = -2$ $M_5(t, u, v) = -4.66$ $M_{54} := -4.66$

$v := z_2 + z_1 = -1.96$

$t := x_2 - x_1 + L_1 = 2$

$u := y_2 - y_1 + L_2 = 6$ $M_5(t, u, v) = 17.631$ $M_{55} := 17.631$

$t := x_2 - x_1 + L_1 = 2$

$u := y_2 - y_1 - L_2 = -2$ $M_5(t, u, v) = 3.471$ $M_{56} := 3.471$

$t := x_2 - x_1 - L_1 = -6$

$u := y_2 - y_1 + L_2 = 6$ $M_5(t, u, v) = -10.862$ $M_{57} := -10.862$

$t := x_2 - x_1 - L_1 = -6$

$u := y_2 - y_1 - L_2 = -2$ $M_5(t, u, v) = -8.635$ $M_{58} := -8.635$

$M_5 := M_{51} - M_{52} - M_{53} + M_{54} + M_{55} - M_{56} - M_{57} + M_{58} = 41.442$

$r_{9.10} := \dfrac{\rho}{4 \times \pi \times L_1 \times L_2} \times M_5 = 41.2231$ ohms   mutual resistance between segments 9 and 10

With the help of computations, it has been found that the mutual resistance between segments 9 and 10 is equal to the mutual resistance between segments 1 and 2.

Self resistance of segment 1, center of segment 1 (−2, −4, −1)

$x_1 := -2$ $y_1 := -4$ $z_1 := -1$ $L := 4$ $a := 0.00526$

$\rho := 200$ ohm–meter

See Eqs. (9.29) and (9.30)

$M_S(t, u) := t \cdot \ln\left[\left(t + \sqrt{t^2 + u^2}\right) - \sqrt{t^2 + u^2}\right]$ Formula

Ground Mat Design with Nonuniform Current Distribution 195

$$M_S(t, u) = M_{S1}(L, a) + M_{S2}(-L, a) - 2a + M_{S3}(L, |z_1| \cdot \sqrt{2}) + M_{S4}(-L, |z_1| \cdot \sqrt{2})$$
$$- 2|z_1| \cdot \sqrt{2}$$

$$M_S := (M_{S1} + M_{S2} - 2a + M_{S3} + M_{S4} - 2|z_1| \cdot \sqrt{2})$$

$$r_{1.1} := \frac{\rho}{4 \cdot \pi \cdot L^2} \cdot M_S$$

$u := a = 0.00526$

$t := L = 4$ $M_S(t, u) = 4.3178$ $M_{S1} := 4.3178$

$t := -L = -4$ $M_S(t, u) = 46.2988$ $M_{S2} := 46.2988$

$u := |z_1| \cdot \sqrt{2}$

$t := L = 4$ $M_S(t, u) = 4.1946$ $M_{S3} := 4.1946$

$t := -L = -4$ $M_S(t, u) = 1.4221$ $M_{S4} := 1.4221$

$$M_S := M_{S1} + M_{S2} - 2a + M_{S3} + M_{S4} - 2|z_1| \cdot \sqrt{2} = 53.394$$

$$r_{1.1} := \frac{\rho}{4 \cdot \pi \cdot L^2} \cdot M_S = 53.112 \text{ ohms} \qquad \text{self resistance of segment 1}$$

Self resistance of segment 9, center of segment 9 (0, –2, –1)

$x_1 := 0$ $y_1 := -2$ $z_1 := -1$ $L := 4$ $a := 0.00526$

For x-directed and y-directed segments, the self resistance is calculated using the same formulas. Also, segment 9 has same length, same radius, and same depth of burial as segment 1, therefore: $r_{9.9} = r_{1.1}$

$r_{9.9} = 53.112$ ohms   Self resistance of segment 9

Matrix element computation:

$$R_{11} = r_{1.1} + r_{1.2} + r_{1.3} + r_{1.4} + r_{1.5} + r_{1.6} + r_{1.7} + r_{1.8}$$

$$R_{22} = r_{9.9} + r_{9.10} + r_{9.11} + r_{9.12}$$

$$R_{12} = r_{1.9} + r_{1.10} + r_{1.11} + r_{1.12}$$

$$R_{21} = r_{9.1} + r_{9.2} + r_{9.3} + r_{9.4} + r_{9.5} + r_{9.6} + r_{9.7} + r_{9.8}$$

Chapter Ten

where

$r_{1.4} = 14.673$　　$r_{1.6} = 15.477$　　$r_{9.11} = 100.685$

$r_{1.5} = 13.633$　　$r_{1.8} = 63.821$

$r_{9.7} = r_{9.2} = r_{1.10} = 25.209$　　$r_{9.8} = r_{9.1} = r_{1.9} = r_{1.2} = 41.223$　　$r_{1.11} = r_{1.3} = 20.814$

$r_{9.6} = r_{9.3} = r_{1.12} = 21.722$　　$r_{9.10} = r_{9.1} = r_{1.9} + r_{1.2} = 41.223$　　$r_{1.1} = r_{9.9} = 53.112$

$r_{9.5} = r_{9.4} = r_{1.7} = 20.816$　　$r_{9.12} = r_{9.10} = r_{9.1} = r_{1.9} = r_{1.2}$

　　　　　　　　　　　　　　　　$= 41.223$

$R_{11} = 53.112 + 41.223 + 20.814 + 14.673 + 13.633 + 15.477 + 20.816 + 63.821 \rightarrow R_{11}$
$= 243.569$

$R_{22} = 53.112 + 41.223 + 100.685 + 41.223 \rightarrow R_{22} = 236.243$

$R_{1.2} = 41.223 + 25.209 + 20.814 + 21.722 \rightarrow R_{1.2} = 108.968$

$R_{21} = 41.223 + 25.209 + 21.722 + 20.816 + 20.816 + 21.722 + 25.209 + 41.223 \rightarrow R_{21}$
$= 217.94$

$$R_{1.1} := 243.569 \qquad R_{1.2} := 108.968$$

$$R_{2.1} := 217.94 \qquad R_{2.2} := 236.243$$

The bus admittance matrix is:

$$\begin{pmatrix} R_{1.1} & R_{1.2} \\ R_{2.1} & R_{2.2} \end{pmatrix}^{-1} = \begin{pmatrix} 6.991 \times 10^{-3} & -3.225 \times 10^{-3} \\ -6.449 \times 10^{-3} & 7.208 \times 10^{-3} \end{pmatrix} \quad Y_{bus} := \begin{pmatrix} 6.991 \times 10^{-3} & -3.225 \times 10^{-3} \\ -6.449 \times 10^{-3} & 7.208 \times 10^{-3} \end{pmatrix}$$

$$\begin{pmatrix} I_1 \\ I_2 \end{pmatrix} := \begin{pmatrix} 6.991 \times 10^{-3} & -3.225 \times 10^{-3} \\ -6.449 \times 10^{-3} & 7.208 \times 10^{-3} \end{pmatrix} \cdot \begin{pmatrix} 1000 \\ 1000 \end{pmatrix}$$

$$\begin{pmatrix} 6.991 \times 10^{-3} & -3.225 \times 10^{-3} \\ -6.449 \times 10^{-3} & 7.208 \times 10^{-3} \end{pmatrix} \cdot \begin{pmatrix} 1000 \\ 1000 \end{pmatrix} = \begin{pmatrix} 3.766 \\ 0.759 \end{pmatrix}$$

$I_1 := 3.776$　　Type I amperes per thousand volts in grid conductors

$I_2 := 0.759$　　Type II amperes per thousand volts in grid conductors

Ground Mat Design with Nonuniform Current Distribution

Total grid leakage current per thousand volts:

$$8 \cdot I_1 + 4 \cdot I_2 = 33.244$$

Ground grid resistance to remote earth:

$$R_g := \frac{1000}{33.244} = 30.081 \text{ ohms}$$

Leakage currents type I_1 and I_2 as well as the total grid leakage current I are proportional to the potential of the grid conductors, and the grid resistance to remote earth is constant and independent of the potential of the grid conductors.

Computation of the Mesh Voltage

The mesh voltage is the maximum potential difference between the mesh conductor and *a point on the earth surface* within the projection of the mesh boundary on the earth surface. For convenience, it is assumed that the geometric center of the mesh projected on the earth surface is the point that maximizes the potential difference. The grid shown in Fig. 10-2 has only four meshes, and all are corner meshes, so any one of them could be selected for the computation of the maximum mesh voltage. Selecting the mesh bound by segments 7, 8, 9, and 12 and assuming that the grid conductor's potential is 1000 volts. Besides, from the example solved in section 9.3, we know that the potential produced at point P per unit ampere of leakage current from any of these segments is 21.4063 volts. We also know that the current leaking out from segments 7 and 8 is 4.013 amperes each. The reader should not confuse the voltage of the mesh conductors with the potential produced at the earth's surface by the current leaking out from the mesh segments.

$$V_{P7} := 4.013 \times 21.4063 = 85.903 \qquad V_{P8} := 4.013 \times 21.4063 = 85.903$$

These are earth surface potential at point P per 1000 volts of mesh conductors potential.

Segments 7 and 8 produce the same potential at P(2, −2, 0) per unit ampere (see Fig. 10-3) because they are equal in length and are the same distance from point P. Therefore it is valid to conclude that the other two segments that surround the mesh produce also the same potential per unit ampere, because they are all of the same length and are at the same distance. The current leaking out from segments 9 and 12 is 0.531 amperes each. So

$$V_{P9} := 0.531 \times 21.4063 = 11.367 \qquad V_{P12} := 0.531 \times 21.4063 = 11.367$$

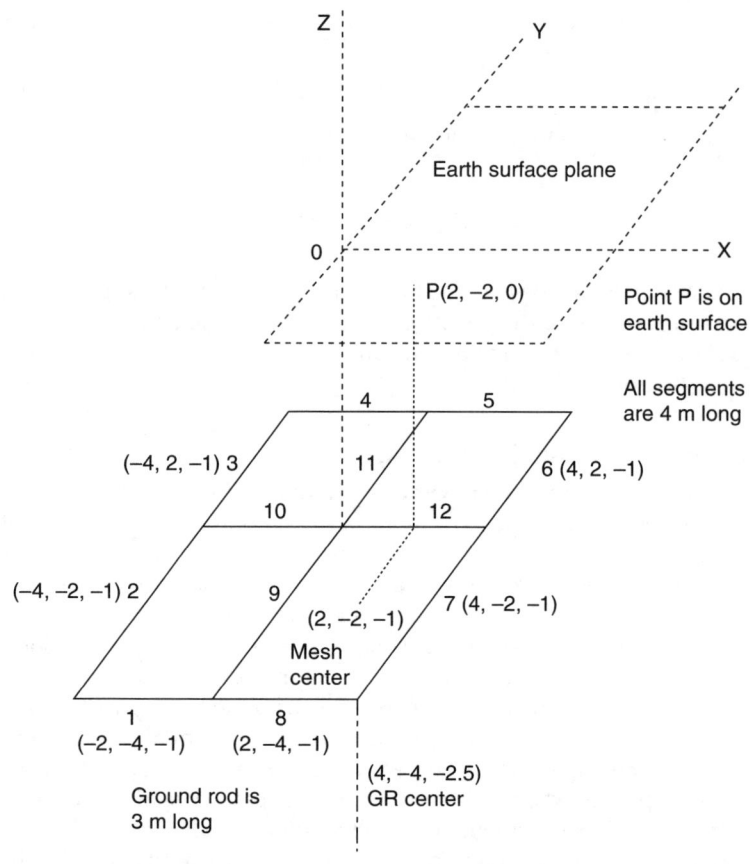

FIGURE 10-3 Square grid buried 1 meter below the earth surface.

The effect of the remaining eight segments (1, 2, 3, 4, 5, 6, 10, 11) must be calculated.

$\rho := 200 \quad L_1 := 4$
$x := 2 \quad y := -2 \quad z := 0 \quad$ geometric center of mesh projected on earth surface

Center point of x-directed segment 1:

$x_1 := -2 \quad y_1 := -4 \quad z_1 := -1 \quad I_1 := 4.013$

$M_1(t, u) := \ln\left(t + \sqrt{t^2 + u^2}\right) \quad$ Formula

$H_{xa} := \sqrt{(y-y_1)^2 + (z+z_1)^2} \quad H_{xa} = 2.236 \quad z + z_1 = -1$

$H_{xb} := \sqrt{(y-y_1)^2 + (z-z_1)^2} \quad H_{xb} = 2.236 \quad z - z_1 = 1$

Ground Mat Design with Nonuniform Current Distribution

From Eq. (9.11) for x-directed segments, we have

$$M_1 = M_{11} - M_{12} + M_{13} - M_{14} \qquad V_{P.1} = \frac{\rho \times I_1}{4 \times \pi \times L_1} \times M_1$$

$$M_1(t, u) = M_{11}(x - x_1 + L_1, H_{xb}) - M_{12}(x - x_1 - L_1, H_{xb}) + M_{13}(x - x_1 + L_1, H_{xa})$$
$$\qquad - M_{14}(x - x_1 - L_1, H_{xa})$$

$u := H_{xb} = 2.236$

$t := x - x_1 + L_1 = 8 \qquad M_1(t, u) = 2.792 \qquad M_{1.1} := 2.792$

$t := x - x_1 - L_1 = 0 \qquad M_1(t, u) = 0.805 \qquad M_{1.2} := 0.805$

$u := H_{xa} = 2.236$

$t := x - x_1 + L_1 = 8 \qquad M_1(t, u) = 2.792 \qquad M_{1.3} := 2.792$

$t := x - x_1 - L_1 = 0 \qquad M_1(t, u) = 0.805 \qquad M_{1.4} := 0.805$

$$M_1 = M_{1.1} - M_{1.2} + M_{1.3} - M_{1.4} = 3.974 \qquad V_{P1} := \frac{\rho \times I_1}{4 \times \pi \times L_1} \times M_1 = 63.454$$

Segment 10:

Type 2 segment, M_1 for this segment is identical to the one for segment 1.

$I_1 := 0.513$

$$V_{P10} := \frac{\rho \times I_1}{4 \times \pi \times L_1} \times M_1 = 8.396$$

For y-directed segment 2:

Type I segment, whose center point coordinates and leaking current are given below.

$x_1 := -4 \qquad y_1 := -2 \qquad z_1 := -1 \qquad I_1 := 4.013$

$H_{ya} := \sqrt{(x - x_1)^2 + (z + z_1)^2} = 6.083 \qquad z + z_1 = -1$

$H_{yb} := \sqrt{(x - x_1)^2 + (z - z_1)^2} = 6.083 \qquad z - z_1 = 1$

$M_1(t, u) := \ln\left(t + \sqrt{t^2 + u^2}\right) \qquad$ Formula

From Eq. (9.12) for y-directed segments, we have

$$M_1 = M_{11}(y - y_1 + L_1, H_{yb}) - M_{12}(y - y_1 - L_1, H_{yb}) + M_{13}(y - y_1 + L_1, H_{ya})$$
$$\qquad - M_{14}(y - y_1 - L_1, H_{ya})$$

$$M_1 = M_{1.1} - M_{1.2} + M_{1.3} - M_{1.4} \qquad V_{P2} = \frac{\rho \times I_1}{4 \times \pi \times L_1} \times M_1$$

$u := H_{yb} = 6.083$

$t := y - y_1 + L_1 = 4$ $\quad M_1(t, u) = 2.423 \quad M_{1.1} := 2.423$

$t := y - y_1 - L_1 = -4$ $\quad M_1(t, u) = 1.188 \quad M_{1.2} := 1.188$

$u := H_{ya} = 6.083$

$t := y - y_1 + L_1 = 4$ $\quad M_1(t, u) = 2.423 \quad M_{1.3} := 2.423$

$t := y - y_1 - L_1 = -4$ $\quad M_1(t, u) = 1.188 \quad M_{1.4} := 1.188$

$M_1 := M_{1.1} - M_{1.2} + M_{1.3} - M_{1.4} = 2.47 \qquad V_{P2} := \dfrac{\rho \times I_1}{4 \times \pi \times L_1} \times M_1 = 39.439$

For y-directed segment 3:

Type 1 segment whose center point coordinates and leaking current are given below.

$x_1 := -4 \quad y_1 := 2 \quad z_1 := -1 \quad I_1 := 4.013$

$H_{ya} := \sqrt{(x - x_1)^2 + (z + z_1)^2} = 6.083$

$H_{yb} := \sqrt{(x - x_1)^2 + (z - z_1)^2} = 6.083 \qquad M_1(t, u) := \ln\left(t + \sqrt{t^2 + u^2}\right) \qquad$ Formula

From Eq. (9.12) for y-directed segments, we have

$u := H_{yb} = 6.083$

$t := y - y_1 + L_1 = 0$ $\quad M_1(t, u) = 1.805 \quad M_{1.1} := 1.805$

$t := y - y_1 - L_1 = -8$ $\quad M_1(t, u) = 0.718 \quad M_{1.2} := 0.718$

$u := H_{ya} = 6.083$

$t := y - y_1 + L_1 = 0$ $\quad M_1(t, u) = 1.805 \quad M_{1.3} := 1.805$

$t := y - y_1 - L_1 = -8$ $\quad M_1(t, u) = 0.718 \quad M_{1.4} := 0.718$

$M_1 := M_{1.1} - M_{1.2} + M_{1.3} - M_{1.4} = 2.174 \qquad V_{P3} := \dfrac{\rho \times I_1}{4 \times \pi \times L_1} \times M_1 = 34.713$

Center point of x-directed segment 4:

$x_1 := -2 \quad y_1 := 4 \quad z_1 := -1 \quad I_1 := 4.013$

$H_{xa} := \sqrt{(y - y_1)^2 + (z + z_1)^2} = 6.083$

$H_{xb} := \sqrt{(y - y_1)^2 + (z - z_1)^2} = 6.083 \qquad M_1(t, u) := \ln\left(t + \sqrt{t^2 + u^2}\right) \qquad$ Formula

Ground Mat Design with Nonuniform Current Distribution

$u := H_{xb} = 6.083$

$t := x - x_1 + L_1 = 8$ $M_1(t, u) = 2.893$ $M_{1.1} := 2.893$

$t := x - x_1 - L_1 = 0$ $M_1(t, u) = 1.805$ $M_{1.2} := 1.805$

$u := H_{xa} = 6.083$

$t := x - x_1 + L_1 = 8$ $M_1(t, u) = 2.893$ $M_{1.3} := 2.893$

$t := x - x_1 - L_1 = 0$ $M_1(t, u) = 1.805$ $M_{1.4} := 1.805$

$M_1 := M_{1.1} - M_{1.2} + M_{1.3} - M_{1.4} = 2.176$ $V_{P4} := \dfrac{\rho \times I_1}{4 \times \pi \times L_1} \times M_1 = 34.745$

Center point of x-directed segment 5:

$x_1 := 2$ $y_1 := 4$ $z_1 := -1$ $I_1 := 4.013$

$H_{xa} := \sqrt{(y - y_1)^2 + (z + z_1)^2} = 6.083$

$H_{xb} := \sqrt{(y - y_1)^2 + (z - z_1)^2} = 6.083$ $M_1(t, u) := \ln(t + \sqrt{t^2 + u^2})$ Formula

$u := H_{xb} = 6.083$

$t := x - x_1 + L_1 = 4$ $M_1(t, u) = 2.423$ $M_{1.1} := 2.423$

$t := x - x_1 - L_1 = -4$ $M_1(t, u) = 1.188$ $M_{1.2} := 1.188$

$u := H_{xa} = 6.083$

$t := x - x_1 + L_1 = 4$ $M_1(t, u) = 2.423$ $M_{1.3} := 2.423$

$t := x - x_1 - L_1 = -4$ $M_1(t, u) = 1.188$ $M_{1.4} := 1.188$

$M_1 := M_{1.1} - M_{1.2} + M_{1.3} - M_{1.4} = 2.47$ $V_{P5} := \dfrac{\rho \times I_1}{4 \times \pi \times L_1} \times M_1 = 39.439$

Center point of y-directed segment 6:

$x_1 := 4$ $y_1 := 2$ $z_1 := -1$ $I_1 := 4.013$

$H_{ya} := \sqrt{(x - x_1)^2 + (z + z_1)^2} = 2.236$

$H_{yb} := \sqrt{(x - x_1)^2 + (z - z_1)^2} = 2.236$ $M_1(t, u) := \ln(t + \sqrt{t^2 + u^2})$ Formula

$u := H_{yb} = 2.236$

$t := y - y_1 + L_1 = 0$ $M_1(t, u) = 0.805$ $M_{1.1} := 0.805$

$t := y - y_1 - L_1 = -8$ $M_1(t, u) = -1.182$ $M_{1.2} := -1.182$

$u := H_{ya} = 2.236$

$t := y - y_1 + L_1 = 0$ $M_1(t, u) = 0.805$ $M_{1.3} := 0.805$

$t := y - y_1 - L_1 = -8$ $M_1(t, u) = -1.182$ $M_{1.4} := -1.182$

$M_1 := M_{1.1} - M_{1.2} + M_{1.3} - M_{1.4} = 3.974$ $V_{P6} := \dfrac{\rho \times I_1}{4 \times \pi \times L_1} \times M_1 = 63.454$

For segment 11 the current is $I_1 = 0.531$.

Segments 6 and 11 have symmetrical position with reference to point P on the earth's surface. Therefore M_1 is the same for both segments. See Figs. 10-2 and 10-3.

$M_1 := 3.974$

$V_{P11} := \dfrac{\rho \times I_1}{4 \times \pi \times L_1} \times M_1 = 8.396$ $V_{P11} := 8.396$

The total potential at point P(2, –2, 0) is

$V_{P1} + V_{P2} + V_{P3} + V_{P4} + V_{P5} + V_{P6} + V_{P7} + V_{P8} + V_{P9} + V_{P10} + V_{P11} + V_{P12} = 486.576$

Mesh voltage:

$V_{mesh} := 1000 - 486.576 = 513.424$ volts

$\rho_s := 3000$ ohm-meters   surface layer resistivity

$t := 1$ second   fault duration

$E_{touch} := \dfrac{116 + 0.17 \times \rho_s}{\sqrt{t}} = 626$ tolerable touch voltage $626 > 513$

$E_{let\text{-}go} := (1000 + 1.5 \times \rho_s) \times 0.009 = 49.5$

Any person submitted to this mesh voltage (513 volts) will become frozen, incapable of releasing the object grasped by his or her hand. The protecting system must clear the fault in less than 1 second.

$V_{P7} + V_{P8} + V_{P9} + V_{P12} = 194.54$ $\dfrac{194.54}{486.576} = 0.4$

Ground Mat Design with Nonuniform Current Distribution

Segment Number	Segment Type	Leakage Current A/1000 Volts	Potential at P Volts
1	I	4.013	63.454
2	I	4.013	39.439
3	I	4.013	34.713
4	I	4.013	34.745
5	I	4.013	39.439
6	I	4.013	63.454
7	I	4.013	85.903
8	I	4.013	85.903
9	II	0.531	11.367
10	II	0.531	8.396
11	II	0.531	8.396
12	II	0.531	11.367

TABLE 10-1 Potential Computations at Point P, Fig. 10-3

The contribution to the total potential at the center point P(2, −2, 0) on the earth's surgface from the segments surrounding the projection of the center point is 40 percent. The remaining eight segments contribute 60 percent.

$$486.576 - 194.54 = 292.036 \qquad \frac{292.036}{486.576} = 0.6$$

By geometric symmetry the mesh voltages of the other three meshes are the same.

Table 10-1 summarizes the result of the computations made to determine the potential at point P of Fig. 10-3.

10.4 Ground Grid Buried in Top Layer of Two-Layer Earth Model

The two-layer model is a better representation of the real ground system, but it greatly complicates the computations, because the two-layer models generate an infinite number of images that become weaker with increasing values of the z-coordinate. Figure 10-4 illustrates the case of two x-directed segments buried in the top layer of a two-layer model where the origin of all the coordinates is on the earth surface and the bottom layer is assumed to be of infinite depth. It is common practice to neglect the superficial layer of crushed stone and backfill material resting on top of the ground grid. When I_1 ampere

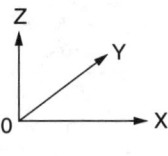

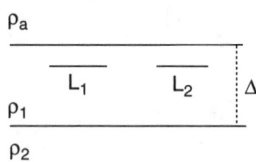

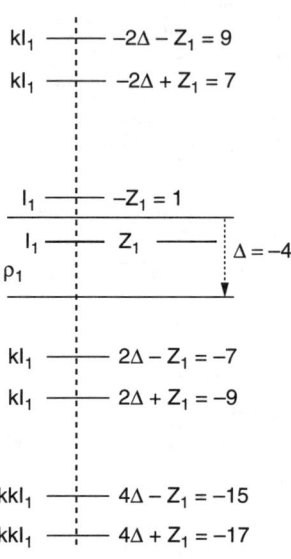

The origin is on earth surface between both segments.

$L_1 = L_2$ segments
$Z_1 = Z_2 = -1$
Δ = first layer thickness
k = reflection coefficient
ρ_a = air resistivity, very large
ρ_1 = first layer resistivity
ρ_2 = second layer resistivity

(a) Real system

(b) Pseudo-equivalent system

Figure 10-4 Two x-directed segments buried in the top layer of a two-layer model.

leaks from segment L_1, segment L_2 is affected not only by segment L_1, but also by its image generated at the boundary between the first layer and the air, and also leaking I_1 ampere. Furthermore, segment L_1 and its image are both reflected in the boundary between layers 1 and 2. The current leaking from the images formed by these reflections at the boundary between layers has a value of kI_1 where k is the reflection coefficient, defined as

$k = \dfrac{\rho_2 - \rho_1}{\rho_2 + \rho_1}$ If $\rho_2 > \rho_1$, then k is positive

If $\rho_2 < \rho_1$ then k is negative

The reflections are unlimited, but consecutive reflections make them weaker when their number increases. Furthermore, the z-coordinate increases by 2Δ with every reflection, so the thicker the top layer, the

Ground Mat Design with Nonuniform Current Distribution

smaller the effect of the image on segment 2. It is a good practice to neglect all the images, except the first one, when the top layer is thicker than 4 meters.

The images' z coordinates on the positive side are

$$-z_1, -2 \times \Delta + z_1, -2 \times \Delta - z_1, \ldots$$

The images' z coordinates on the negative side are

$$z_1, 2 \times \Delta - z_1, 2 \times \Delta + z_1, 4 \times \Delta - z_1, 4 \times \Delta + z_1, \ldots$$

The procedure described in Sec. 9.4 is valid for the two-layer earth model, but it must be generalized to the case of two series of images, one for each boundary. All the images are in ρ_1 territory, because segments L_1 and L_2 are both inside the ρ_1 layer. The necessary thing to do is to expand Eq. (9.15) to include the additional images, taking into consideration that the coordinates of the additional images are the same as those of the original segment except for the z coordinate, which grows with every reflection in both boundaries. Furthermore, the current that leaks from the additional images becomes smaller by the factor k, k^2, ... Equation (9.15) is written for the original segment plus one image.

Expanding it to include four more images, the closer ones to segment L_2 are

$-2\Delta - z_1$	$-2\Delta + z_1$	$2\Delta - z_1$	$2\Delta + z_1$	all with current kI_1
9	7	-7	-9	vertical distances to segment 2 in meters

$$\rho_1 := 100 \quad \rho_2 := 200 \quad L_1 := 4 \quad L_2 := 4 \quad k := \frac{\rho_2 - \rho_1}{\rho_2 + \rho_1} = 0.333$$

$$k^2 = 0.111$$

Segment images with k^2 factors (In Fig. 10-4, k^2 is represented by kk.), in addition to carrying a very small fraction of the original segment's current, are far away from segment 2. They should be neglected. Each of the additional M terms includes one of the following u variables.

$$u_{xc} = \sqrt{(y_2 - y_1)^2 + (z_2 - 2 \cdot \Delta - z_1)^2} \quad u_{xd} = \sqrt{(y_2 - y_1)^2 + (z_2 - 2 \cdot \Delta + z_1)^2}$$

$$u_{xe} = \sqrt{(y_2 - y_1)^2 + (z_2 + 2 \cdot \Delta - z_1)^2} \quad u_{xf} = \sqrt{(y_2 - y_1)^2 + (z_2 + 2 \cdot \Delta + z_1)^2}$$

Sixteen more M terms must be added to Eq. (9.15) to include the effect of the four images added. These terms are of the type

$$M_{29} = (x_2 - x_1 + L_1 + L_2, u_{xc})$$

Figure 10-4 is only an illustration and is not to scale.

10.5 Ground Grid Buried in Bottom Layer of Two-Layer Earth Model

The ground grid is seldom buried more than 1 meter deep. Therefore it is very improbable that in a real case the ground grid was laid buried in the bottom layer. If the first layer is very shallow, the grid designer always has the choice of installing the grid 0.5 meter deep. However, what it is more common is the case of ground rods buried in the first layer, but penetrating all the way down to the second layer.

These cumbersome cases are beyond the scope of this book. However, the procedure discussed in Secs. 10.2 and 10.3 is still valid for these cases, but it would require a lot more monotonous and repetitive computations because it must include the images generated by the boundary reflections. The reader should understand that the mutual resistance affecting a particular segment includes the contribution of all other segments whether they are x-directed, y-directed, or z-directed. Figure 10-5 illustrates the case of two x-directed segments

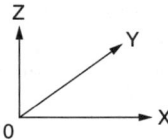

The origin is on earth surface between both segments.

$Z_1 = Z_2 = Z = -5$
Δ = first layer thickness = -4
ρ_a = air resistivity, very large
ρ_1 = first layer resistivity
ρ_2 = second layer resistivity
k = reflection coefficient
$k = 0.3333$

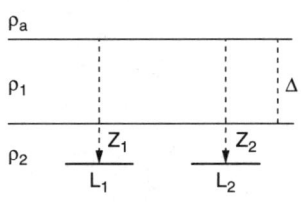

(a) Real system

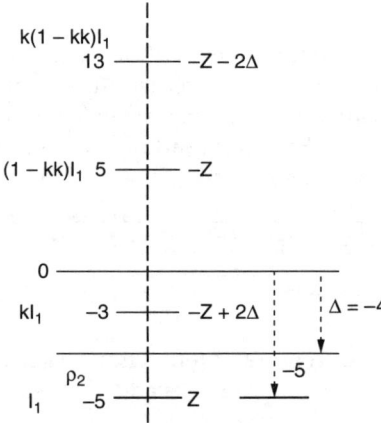

(b) Pseudo-equivalent system

Figure 10-5 Two x-directed segments buried in the bottom layer.

buried in the bottom layer of a two-layer earth model. The first image of L_1 is generated by reflection on the bottom surface of the boundary between layers, and because this reflection does not hit any other reflecting surface (the second layer is infinitely deep) it does not generate any other image; consequently there are no images below segment L_1. The second image of L_1 is generated by reflection on the earth surface. Next, the second image is reflected at the boundary between layers and reflected again at the earth surface, generating a third image. This process continues indefinitely, generating a series of increasing weaker images that are separated by z increments of 2Δ. The leakage currents of the images are indicated in Fig. 10-5.

Bibliography

Anderson, P. M., and A. A. Fouad, *Power System Control and Stability*, Iowa State University Press, 1977.
Bayliss, C. R., *Transmission and Distribution Electrical Engineering*, Butterworth-Heinemann, 1996.
Byerly, R. T., and E. W. Kimbark, *Stability of Large Electric Power Systems*, IEEE Press, 1974.
Grainer, J. J., and W. D. Stevenson, Jr., *Power System Analysis*, McGraw-Hill, 1994.
Grigsby, L. L., *Electrical Power Generation: Transmission and Distribution*, CRC Press, Taylor and Francis Group, 2007.
Grigsby, L. L., *Power System Stability and Control*, CRC Press, Taylor and Francis Group, 2007.
IEEE Guide for Safety in AC Substation Grounding, Std. 80-2000, Institute of Electrical and Electronics Engineers, 2000.
Kuffel, E., W. S. Zaengl, and J. Kuffel, *High Voltage Engineering Fundamentals*, 2d ed., Butterworth-Heinemann, 2000.
Machowski, J., J. W. Bialek, and J. R. Bumby, *Power System Dynamics and Stability*, John Wiley & Sons, 1997.
Machowski, J., J. W. Bialek, and J. R. Bumby, *Power System Dynamics: Stability and Control*, 2d ed., John Wiley & Sons, 2008.
McDonald, J. D., *Electric Power Engineering Handbook*, 2d ed., CRC Press, Taylor and Francis Group, 2007.
Moin, P., *Fundamentals of Engineering Numerical Analysis*, 2d ed., Cambridge University Press, 2010.
Sakis Meliopoulos, A. P., *Power System Grounding and Transients—An Introduction*, Marcel Dekker, 1988.
Stevenson, W. D. Jr., *Elements of Power System Analysis*, 3d ed., McGraw-Hill, 1975.

Index

Note: Page numbers followed by *f* denote figures; page numbers followed by *t* denote tables.

A

AC systems. *See* Alternating-current systems
Admittance:
  bus admittance matrix, 73–74
  bus matrix, 76, 196–197
  equivalent, electrical loads converting to, 71–72
  network, 75
  one-line diagram of 230 kV, 67*f*
  one-line diagram of 300 MVA, 67*f*
  one-phase diagram with loads, generators, and transformers, 67*f*
  power system per-unit equation, 46
  square bus matrix, 74, 76
Admittance matrix, 45. *See also* Bus admittance matrix
  per-unit diagram for, 48*f*
  per-unit equation for, 46
Air gap, 17–18
Alternating-current (AC) systems, 1. *See also* High-voltage AC capacitors
Angular momentum M, 28, 57

B

Balance of power computations, 82
Buried grid in substation grounding, design approach, 127–128

Bus admittance matrix, 73
  computation, 196–197
  of pretransient network, 76
  sparse matrix and, 74
Bus networks, generator-infinity, 32, 39–40, 49–50, 50*f*, 55*f*
Bus voltages before completion of first iteration, 81–82

C

Capacitor:
  coupling, 104
  grading, 104
  high-speed, 40
  high-voltage AC, 103–125
  instantaneous voltage across, 116
  power factor correction, 104
  steady-state equations, 105
  TRVs, 104
  voltage transformers, 104
Capacitor bank, 1
  charging current after fully discharged, 119
  charging current during first 200 ms, 122*f*
  charging current during the first 4 seconds, 124*f*
  charging current during the first second, 123*f*
  discharging after single line-to-ground fault, 115, 118*f*
  LN voltage oscillations, 124*f*
  numerical solver, 119–123

Index

Capacitor bank (*Cont.*):
  parameters per phase, 115
  in picofarad range, 118
  series, 40, 109–110, 109f
  series-connected, 108–110
  shunt-connected, 110–112, 111f
Capacitor connections:
  capacitance from $KVAR_c$ to picofarads, 106, 114–115
  in parallel, 106
  in series, 105–106
Cartesian coordinates, Laplacian operator in, 162
CCs. *See* Coupling capacitors
Charging current:
  during the first 4 seconds in capacitor bank, 124f
  during first 200 ms in capacitor bank, 122f
  during the first second in capacitor bank, 123f
  after fully discharged in capacitor bank, 119
Classical model:
  generator and transformer, 73f
  of multimachine power systems, 43
  power systems stability, 18–20, 19f
  single machine, 42–43
  synchronous generators, 18–20
Coherent generators, 41
Coherent machines, 41–42
Complex power equation, 22
Components in parallel, 9–10
  equation, 9
  rule, 9
Components in series, 8–9
  equivalent circuit of a circuit, 8f
  rule, 8f, 9
Copper cables, 144–145
  data, 147t
  maximum allowed temperatures, 146
Coupling capacitors (CCs), 104

D

Dalziel equation, 131
Damping power in multimachine power systems, 42

DC offset during short circuit conditions, 20–21
Delta-wye conversion:
  equation, 10
  examples, 11–12
  during fault conditions, 51–52, 52f
  series reactances addition, 90
  short circuit MVA combination rules, 10–12, 10f
Dielectric insulation material, 103
Differentiable scalar function, Laplace homogeneous equation for, 162
Distribution:
  grid current, during fault to ground, 177–180, 178f
  leakage current, 177–180, 178f
  nonuniform current in square ground grid, 180–203
  three-phase balanced circuit for, 1
Double line-to-ground short circuit, 39

E

Earth:
  ground mat resistance to, 141f, 142–143
  point inside, 163–167
  point on surface, 197
  single rod electrodes resistance to, 141f
  square ground grid buried below, 163–164, 164f, 198f
Earth model:
  one-layer, 159
  two-layer, 159, 203–207, 204f, 206f
Electric currents, dangerous:
  background, 131
  Dalziel equation for, 131
  duration and path, 133–137
  electrical substation grounding, 137–138
  human injury, 132, 134
  important voltage gradient definitions, 138
  magnitude and frequency, 132

Electric power transmitted:
  after fault, 54, 63, 63f
  before fault, 53
  during fault conditions, 54
  in transient stability problem in simple electrical network, 53–54
Electrical field, 104
Electrical loads:
  equivalent admittances converting from, 71–72
  in multimachine power systems, 43
Electrical networks. *See also* Simple electrical network
  mechanical formulas for transient stability of, 27t
Electrical networks reduction:
  Norton's theorem, 2–4, 3f
  superposition theorem, 2, 4
  Thévenin's theorem, 2, 3f, 4
Electrical quantities, per-unit method of, 4–6, 45, 46, 48f, 55–56, 67f
Electrochemical potential series, 144f
Electromagnetic induction law, 15–16
Electromotive force (emf), 18
Equal-area criterion of stability, 36
  multimachine power systems and, 43
  power flow, 37–39, 38f
Equivalent admittances, electrical loads converting to, 71–72
Equivalent circuit of a circuit with components in series, 8f
Equivalent delta network, during fault condition, 51–52, 52f
E_{step}. *See* Tolerable step voltage
E_{touch}. *See* Tolerable touch voltage

F

Fault:
  cleared, G1 swing equation after, 96
  cleared, G2/G3 swing equation after, 96

Fault (*Cont.*):
  cleared in simple electrical network, 49, 60
  conditions, power unbalance during, 36
  electric power transmitted after, 54, 63, 63f
  electric power transmitted before, 53
  electric power transmitted during, 54
  equal-area criterion of stability of, 36–39
  at F1, network configuration during, 89–96
  GMPR and, 160
  to ground, grid current distribution during, 177–180, 178f
  in multimachine power systems, 43
  network with three-phase, 36f
  response by power system, 41
Fault conditions:
  G1 swing equation during, 95–97
  G2/G3 swing equation during, 95, 97–100
  matrix of power angles during, 99f
First iteration in G-S method, 78–83
  bus voltages before completion of, 81–82
  real and reactive powers after first iteration, 83
  real and reactive powers before completion, 82

G

G-S. *See* Gauss-Seidel method
G1 rotor natural frequency oscillation period, 101–102
G1 swing equation:
  during fault conditions, 95–97
  after fault is cleared, 96
G2/G3 rotor natural frequency oscillation period, 101–102

G2/G3 swing equation:
  during fault conditions, 95, 97–100
  after fault is cleared, 96
G2 initial power angle, 88–89
G3 initial power angle, 89
Gauss-Seidel (G-S) method, 74–75, 77
  first iteration, 78–83
  second iteration of, 84–86
Generated voltage, 20
Generator. *See also* Synchronous generators
  busses, 44
  classical model, 73f
  coherent, 41
  factory ratings, 67f
  maximum electric output before fault, 53
  mechanical rotor angle, 43
  one-phase admittance diagram, 67f
  per-unit values converted to 230-kV and 300-MVA, 68f
  reactance between, 95f
  steady-state operating point of, 33
  terminal voltage, 22
Generator-infinity bus network:
  high-speed circuit breakers use in, 40
  parallel transmission lines use in, 40, 49, 50f
  reactance before fault in, 50
  series capacitor banks use in, 40
  swing equation in, 32
  system voltage increase, 39–40
  transient stability criterion, 55f
  transient stability improvement methods of, 32, 39–40
Generator to motor:
  complex power equation, 22
  equations, 22
  power flow from, 20–25
  stability limits, 24–25
GMPR. *See* Grounded metal potential rise
Gradient control, 148–152

Grading capacitors, 104
Grid conductor:
  reclosing sequence, 146
  size, 145–148
  temperature, 146
Grid conductor segments:
  creation from current leaking to earth, 163–167
  mutual resistance between, 167–173
  two parallel, x-directed line segments, 168–169
Grid current distribution during fault to ground, leakage current distribution, 177–180, 178f
Ground grid:
  buried in bottom layer of two-layer earth model, 206–207, 206f
  buried in substation grounding, 127–128
  buried in top layer of two-layer earth model, 203–205, 204f
  square, buried below earth, 163–164, 164f, 198f
Ground grid preliminary design:
  background, 139–140
  copper cables in, 144–145
  design procedure, 153–155, 153f
  determinations for, 139
  example of, 152–158
  gradient control, 148–152
  grid conductor size, 145–148
  ground mat conductor corrosion, 143–145
  ground mat resistance to earth, 141f, 142–143
  return ground current check, 157–158
  single-rod electrodes, 140–141
  soil layers in, 159
  V_{step} computation, 156–157
Ground mat:
  conductor corrosion, 143–145
  equation, 142
  resistance to earth, 141f, 142–143
  typical, 140f

Ground mat design principles:
  GMPR, 160
  mutual resistance between two conductor segments, 167–173
  one-layer earth model, 159
  penetration depth, 160–161
  point current source potential creation, 161–163
  point inside earth created by current leaking, 163–167
  procedure, 160
  self-resistance, 174–175
  two-layer earth model, 159
Ground mat design with nonuniform current distribution:
  computations in small square grid, 180–203
  grid current distribution during a fault to ground, 177–180
  ground grid buried in bottom layer of two-layer earth model, 206–207, 206f
  ground grid buried in top layer or two-layer earth model, 203–205, 204f
Grounded metal potential rise (GMPR), 160

H

Harmonics, power system introduction of, 19
High-speed capacitor, generator-infinity bus network use of, 40
High-voltage AC capacitors, 103
  AC voltage applied to or removed from RLC series circuit, 112–125
  basic capacitor connections, 105–106, 114–115
  capacitor steady-state equations, 105
  capacitor voltage transformers, 104
  CCs, 104
  classifications, 104
  grading capacitors, 104

High-voltage AC capacitors (*Cont.*):
  power factor correction capacitors, 104
  reactive power compensation, 107–108
  series-connected capacitor banks, 108–110
  shunt-connected capacitor banks, 110–112
  TRVs, 104
Human injury from electric shock:
  muscular control loss, 132
  muscular inhibition, 132
  perception, 132
  reaction or surprise, 132
  severe electric shock, 132
  ventricular fibrillation of heart, 132, 134
Hysteresis loop, 15

I

IEEE Guide for Safety in AC Substation Grounding, IEEE Standard 80, 129
Impedance:
  diagram of power system, 50f
  line, per-unit, 69f
  network diagram, 23f
  per-unit, 6
  shunt line, 22
Inertial constant, 26
  H, 28–29, 35, 41
Initial power angle computation:
  G1, 87–88
  G2, 88–89
  G3, 89
Instantaneous voltage across capacitor, 116
Insulation material, dielectric, 103
Iron core saturation, 12
  electromagnetic induction law, 15–16
  hysteresis loop, 15
  sinusoidal wave equation, 13, 13f
  voltage applied to winding, 13f

Index

K

Kilovolt-amperes reactive (KVAR), 106
Kilovolts (kV), 230, one-line admittance diagram, 67f
KVAR. *See* Kilovolt-amperes reactive
$KVAR_c$, picofarads capacitance from, 106, 114–115

L

Laplace homogeneous equation for differentiable scalar function, 162
Laplacian operator in cartesian coordinates, 162
Leakage current distribution:
  equations, 178–180
  formula for, 180
  geometric symmetry and, 177
  matrix equation, 180
  six currents, 178f
Lenz' law, 19
Let-go voltage, 136f
Line contribution to short circuit MVA, 11
Line impedances, per-unit, 69f
Line-to-ground:
  fault, 115, 118f
  short circuit, 39
Line-to-line short circuit, 39
Line-to-line (LL) voltage, 72
  per unit value of, 5
Line-to-neutral (LN) voltage, 113
  electrical load converting to, 71
  instantaneous midpoint, 116
  oscillations, 124f
  per-unit value of, 5
Linear, two-terminal circuits, Thévenin's theorem and, 2
Linear motion formula, 26
Linear system model, 25
LL voltage. *See* Line-to-line voltage
LN voltage. *See* Line-to-neutral voltage

Load. *See also* Electrical loads
  busses, 44
  connection to network, 19
  flow in multimachine network, 73–86
  one-phase admittance diagram with, 67f
  representation on stability limits, 71
  steady-state stability for changes in small, 34–35
  study, 1
Load flow during normal operation, 73
  first iteration, 78–83
  G-S method, 74–75, 77–78
  second iteration, 84–86
Lossless short transmission line, 113f

M

Magnetomotive force (mmf), 18
MathCad, 32, 120
Matrix:
  admittance, 45, 46, 48f
  bus admittance, 73–74, 76, 196–197
  leakage current distribution equation, 180
  nodal elimination, 73
  of power angles during fault conditions, 99f
  R elements, 182–197
  solution, 61f, 121f, 122f
  sparse, 74
  square bus admittance, 74, 76
  V elements, 180
Matrix R elements:
  bus admittance matrix computation, 196–197
  computation, 195–196
  segments 1 and 2, center of segment 1 (−2, −4, −1), center of segment 2 (−4, −2, −0.96) computation, 188
  segments 1 and 3, center of segment 1 (−2, −4, −1), center of segment 3 (−4, 2, −0.96) computation, 189–190

Matrix R elements (*Cont.*):
  segments 1 and 4, center of segment 1 (−2, −4, −1), center of segment 2 (−2, 4, −1) computation, 183–184
  segments 1 and 5, center of segment 1 (−2, −4, −1), center of segment 2 (2, 4, −1) computation, 184–185
  segments 1 and 6, center of segment 1 (−2, −4, −1), center of segment 6 (4, 2, −1) computation, 190–191
  segments 1 and 7, center of segment 1 (−2, −4, −1), center of segment 7 (4, −2, −0.96) computation, 191–192
  segments 1 and 8, center of segment 1 (−2, −4, −1), center of segment 2 (2, −4, −1) computation, 186–187
  segments 1 and 10, center of segment 1 (−2, −4, −1), center of segment 2 (−2, 0, −1) computation, 185–186
  segments 1 and 12, center of segment 1 (−2, −4, −1), center of segment 2 (2, 0, −1) computation, 186
  segments 4 and 9, center of segment 1 (−2, 4, −1), center of segment 2 (0, −2, −0.96) computation, 192–193
  segments 9 and 10, center of segment 1 (0, −2, −1), center of segment 2 (−2, 0, −0.96) computation, 193–194
  segments 9 and 11, center of 1 0, −2, −1), center of segment 2 (0, 2, −1) computation, 187–188
  self-resistance and mutual resistance in, 182–183
  self-resistance of segment 1, center of segment 1 (−2, 04, −1) computation, 194–195
  self-resistance of segment 9, center of segment (0, −2, −1) computation, 195

Matrix V elements, 180
Maximum real power equation, 22, 24
Mechanical power in multimachine power systems, 42
Megajoules:
  angular momentum expressed in, 57
  rotational energy conversion to, 29–30
Megavolt-amperes (MVA), 28, 29. *See also* Short circuit MVA combination rules
  300, one-line admittance diagram, 67*f*
  equation, 5
  short circuit calculation method, 6–7
  three-phase, 4
Mesh voltage (V_{mesh}), 138, 148, 149, 151*f*
  potential computations, 203*f*
  for y-directed segment, 199–202
Mesh voltage (V_{mesh}) computation, 202–203
  center point of x-directed segment, 198–201
  point on earth surface and, 197
Metal oxide varistor (MOV), 110
mmf. *See* Magnetomotive force
Motor contribution to short circuit MVA, 11
MOV. *See* Metal oxide varistor
Multimachine power systems:
  assumptions in, 42–43
  classical model of, 43
  damping power in, 42
  electrical loads in, 43
  equal-area criterion of stability and, 43
  faults in, 43
  generator mechanical rotor angle in, 43
  mechanical power in, 42
  modeling, 42–43
  single-line diagram for power flow in, 43–44
  single machine classical model in, 42–43
  stability, 40–41

Index

Mutual resistances of segments in matrix R elements, 182–183
MVA. *See* Megavolt-amperes

N

Network:
  admittance, 75
  bus generator-infinity, 39–40, 49–50, 50f, 55f
  configuration during fault at F1, 89–96
  electrical, 27t
  electrical reduction, 2–4
  equivalent delta, 52, 52f
  equivalent symmetrical Π, 20–21, 20f
  fault with three-phase, 36f
  impedance diagram, 23f
  load connection, 19
  multimachine network, transient stability problem in, 65–102
  pretransient, 76
  reduction in transient stability problem in simple electrical network, 50–52
  simple electrical, 49–64, 50f, 51f, 52f
Network reduction techniques, 44–48
  delta-wye transformation formulas, 44
  star-mesh transformation formulas, 44
Nodal elimination matrix, 73
Node elimination method, 44–47
Norton's theorem, 2, 3f
  equation, 4
  two-terminal circuits and, 3
Numerical solver:
  capacitor bank, 119–123
  single vector function, 59–60
  standardization, 58–59
  swing equation, 96–102
  in transient stability problem in simple electrical network, 58–64

O

Ohm's law, 4
One-layer earth model, 159
One-phase admittance diagram with loads, generators, and transformers, 67f
One-phase illustration of a series capacitor bank, 109f
Open-phase mode power system failure, 1
Oscillation:
  LN voltage, 124f
  natural frequency, 35
  period of, G1 rotor natural frequency and, 101–102
  period of, G2/G3 rotor natural frequency and, 101–102
  of power angle, 34
  subharmonic, 35
  voltage during first four seconds, 123–125
Output voltage of synchronous generators, 17

P

P. Laurent grid, 142, 149
Parallel:
  capacitor connections in, 106
  components in, 9–10
  transmission lines, generator-infinity bus network use of, 40, 49, 50f
  two, in X-directed segment of point current source, two parallel, 168–169
  in Y-directed segment of point current source, parallel, 169–170
  in Z-directed segment of point current source, 170
Per-phase calculations, in three-phase balanced circuit, 1
Per-unit impedance, 6
Per-unit quantities:
  230 kV, 67f
  300 MVA, 67f
  in admittance matrix, 45, 46, 48f
  before, during, after fault conditions, 55–56

Per-unit quantities (*Cont.*):
  impedance values, 6
  line-to-line kilovolts, 4
  Ohm's law, 4
  percentage value, 4–5
  three-phase MVA, 4
Perpendicular segments of point current source, 170, 172–173
Phase to ground short circuit:
  current production of, 1
  failure, 1
Phase to phase short circuit failure, 1
Picofarads:
  $KVAR_c$ capacitance to, 106, 114–115
  range, capacitor bank in, 118
Point current source:
  Laplace equation and, 162
  perpendicular segments, 170, 172–173
  potential created by, 161–163
  single-point-source solution, 162
  x-directed segment, 164–165, 168–169, 174, 198–201
  y-directed segment, 164, 166, 169–170, 174, 199–202
  z-directed segment, 165, 167, 170, 175
Point inside earth, 163–167
Point on earth surface, V_{mesh} and, 197
Power. *See also* Reactive power; Real power
  accelerating, power angle and, 61*f*, 63*f*
  balance of, computations, 82
  coefficient synchronization, 32–35
  complex equation, 22
  damping, 42
  factor correction capacitors, 104
  generation, three-phase balanced circuit for, 1
  shaft in steady-state condition, 93
  triangle, 107*f*
  unbalance during fault conditions, 36

Power angle:
  accelerating power and, 61*f*, 63*f*
  before, during, and after fault conditions, 56
  in equal-area criterion of stability, 38–39
  G1 initial, 87–88
  G2 initial, 88–89
  G3 initial, 89
  matrix of during fault conditions, 99*f*
  oscillations of, 34
  in radians, 100*f*
Power flow:
  in equal-area criterion of stability, 37–39, 38*f*
  from generator to motor, 20–25
  in multimachine network, 43–44
  single-line diagram in multimachine power system, 43–44
Power system:
  electrical networks reduction, 2–4
  external subsystem of, 70
  fault response by, 41
  harmonics introduction, 19
  impedance diagram of, 50*f*
  internal subsystem of, 70
  iron core saturation, 12–16, 13*f*, 14*f*
  MVA method of short circuit calculation, 6–7
  open-phase mode failure, 1
  per-unit admittance equation, 46
  per-unit quantities, 4–6, 45, 46, 48*f*, 55–56, 67*f*
  short circuit mode failure, 1
  short circuit MVA combination rules, 7–12
  split to facilitate analysis, 69*f*
  three-phase balanced circuits, 1–2
  time-domain methods of, 66
  transient stable, 65–66
Power systems stability, 17
  classical model, 18–20, 19*f*
  coherent machines, 41–42
  equal-area criterion of stability, 36–39

Power systems stability (Cont.):
  generator-infinity bus network, 32, 39–40
  multimachine power systems modeling, 42–43
  multimachine power systems stability, 40–41
  network reduction techniques, 44–48
  oscillation natural frequency, 35
  power coefficient synchronization, 32–35
  power flow from generator to motor, 20–25
  power flow in multimachine network, 43–44
  rotational dynamics summary, 26–30
  steady-state stability, 25
  swing equation, 30–32
Power transmitted before, during and after fault conditions, in transient stability problem in simple electrical network, 55–56
Pretransient network, bus admittance matrix of, 76

Q

Quadrature axis, 18
Quadrature transient reactance, 19

R

Reactance:
  between generator, 95f
  of network configuration during Fault at 1, 91f
  series, 90
  shunt capacitive, 23
  transient, 19, 20, 50–52, 51f, 52f
Reactive power:
  compensation, 107–108
  after completion of first iteration, 83
  before completion of first iteration, 82
  equation, 22, 24
  estimated per-unit, 69f, 72
  of synchronous generators, 17
Real power:
  after completion of first iteration, 83
  before completion of first iteration, 82
  estimated per-unit, 69f, 72
  maximum, equation, 22, 24
  of synchronous generators, 17
Rkadapt function, 59–62, 62f, 98, 120
Rotational dynamics summary, 26–30
Rotational energy:
  megajoules conversion in, 29–30, 57
  stored at synchronous speed, 93
Rotational motion formula, 26
Rotor:
  angle, generator mechanical, 43
  angular acceleration equation, 31
  angular position of, 30
  angular velocity equation, 30–31
  G1 natural frequency, 101–102
  G2/G3 natural frequency, 101–102
  studies on stability, 19
  of synchronous generator, 17
Round-rotor synchronous generator, air gap in, 17–18
Runge-Kutta method, 120

S

Salient pole synchronous generator:
  air gap in, 17–18
  direct axis of, 18
  direct axis transient reactance in, 19, 20
  quadrature axis, 18
  quadrature transient reactance, 19
Second iteration of G-S method, 84–86

Index 221

Self-resistance:
   matrix R elements as function of, 182–183, 194–195
   of x-directed segment, 174
   of y-directed segment, 174
   of z-directed segment, 175
Separated ground rods, 129
Series:
   capacitor connections in, 105–106
   components in, 8–9, 8f
   electrochemical potential, 144f
   reactance, 90
   RLC circuit, 112–125
Series capacitor banks:
   connected in series with line, 109–110
   generator-infinity bus network use of, 40
   one-phase illustration of, 109f
Series-connected capacitor banks, 108–110
Shaft power in steady-state condition, 93
Short circuit conditions, DC offset during, 20–21
Short circuit failure:
   double line-to-ground, 39
   line-to-ground, 39
   line-to-line, 39
   phase to ground, 1
   phase to phase, 1
   three-phase, 1, 39
   two-phase to ground, 1
Short circuit mode, power system failure in, 1
Short circuit MVA combination rules, 7–12, 11f
   components in parallel, 9–10
   components in series, 8–9, 8f
   delta-wye conversion, 10–12, 10f
   line contribution, 11
   motor contribution, 11
   transformer contribution, 11
   utility contribution, 11

Shunt:
   capacitive reactance equation, 23
   -connected capacitor banks, 110–112, 111f
   line impedances, 22
Simple electrical network:
   after combining transient reactance, 52, 52f
   electric power transmitted in, 53–54
   fault cleared in, 49, 60
   network reduction in transient stability in, 50–52
   numerical solver in transient stability in, 58–64
   power transmitted before, during and after fault conditions in, 55–56
   single-line diagram of, 50f
   stability problem in, 49–50
   swing equation in transient stability in, 56–58
   during three-phase symmetrical short circuit, 51–52, 51f
   transient stability in, 49–64
Single machine classical model, in multimachine power system, 42–43
Single-phase circuit analysis, 1
Single point current source, methods of images applied to, 161f
Single-point-source solution, 162
Single rod electrodes, 140–141
   resistance to earth, 141f
Single vector function, Rkadapt function as, 59–60
Sinusoidal instantaneous midpoint voltage, 115–119
Sinusoidal wave, 13, 13f
Small-disturbance stability, 18
Small load changes, steady-state stability for, 34–35
Soil layers:
   in ground grid designs, 159
   resistivity in, 159–160

Solution matrix, 61f, 121f
  of vector D, 122f
Sparse matrix, 74
Square bus admittance matrix, 74, 76
Square ground grid:
  buried below the earth, 163–164, 164f, 198f
  geometric symmetry, 181
  matrix R elements determination in, 182–197
  nonuniform current distribution in, 180–203
  segment classification, 181–182
  uniformly spaced, 181f
SSR. *See* Subsynchronous resonance
Stability limits:
  of generator to motor, 24–25
  load representation on, 71
Stability problem in simple electrical network, 49–50
Standardization in numerical solver, 58–59
Star-mesh transformation formulas, 44
Stator of synchronous generator, 17
Steady-state stability, 23
  equation, 24
  linear system model and, 25
  for small load changes, 34–35
Steel data, 147t
Step voltage (V_{step}), 138, 148
Stored rotational energy, 28, 57
Subharmonic oscillation, 35
Substation fences, 129–130
Substation grounding:
  assumptions on, 128
  background, 127
  buried grid in, 127–128
  dangerous electrical currents and, 137–138
  E_{touch}, 128
  grid design approach, 127–128
  separated ground rods, 129
  substation fences, 129–130
Subsynchronous resonance (SSR), 108

Superposition theorem, 2, 4
Swing bus voltage magnitude, 81, 83
Swing equation, 25, 34
  G1, 95–97
  G2/G3, 95, 97–100
  in generator-infinity bus network, 32
  numerical solver of, 96–102
  rotor angle velocity equation, 30–31
  rotor angular acceleration equation, 31
  in transient stability problem in simple electrical network, 56–58
Synchronizing power coefficient, 32–35
Synchronous generators:
  classical model, 18–20
  output voltage, 17
  reactive power of, 17
  real power of, 17
  round-rotor, 17
  salient pole, 17
System voltage increase in generator-infinity bus network, 39–40

T

TCR. *See* Thyristor controlled reactor
Thévenin's theorem, 3f
  equation, 4
  linear, two-terminal circuits and, 2
Three-phase balanced circuits, 1–2, 2f
  computations, 5
  for distribution, 1
  equivalent single-line, 2f
  failure, 1
  LL voltage per-unit value, 5
  LN voltage per-unit value, 5
  per-phase calculations in, 1
  for power generation, 1
  for transmission, 1
Three-phase fault network, 36f
Three-phase MVA, 4

Three-phase power system with three generators and three loads, 66f
Three-phase symmetrical short circuit, 39
  simple electrical network during, 51–52, 51f
Thyristor controlled reactor (TCR), 111
  one phase of, 111f
Thyristor switched shunt connected capacitor bank, 111f
Time-domain methods of power system, 66
Tolerable body current, 133f
Tolerable step voltage (E_{step}), 134, 135f, 138, 148
Tolerable touch voltage (E_{touch}), 128, 135, 136f, 138, 148
Touch voltage (V_{touch}), 138, 148
Transferred voltage (V_{trans}), 138, 149
Transformer:
  classical model, 73f
  contribution to short circuit MVA, 11
  per-unit values converted to 230-kV and 300-MVA, 68f
Transient reactance:
  during delta-wye conversion, 51–52, 52f
  direct axis, 19, 20
  before fault in generator-infinity bus network, 50
  simple electrical network after combining, 52, 52f
  during three-phase symmetrical fault, 51–52, 51f
Transient recovery voltage capacitors (TRVs), 104
Transient stability, 18
  generator-infinity bus network improvement methods, 39–40
  mechanical formulas for, 27t
Transient stability problem in multimachine network, 65–102
  converting electrical loads to equivalent admittances, 71–72

Transient stability problem in multimachine network (*Cont.*):
  initial power angle computation, 87–89
  load flow during normal operation, 73–86
  minimum data for study of, 68–71
  network configuration during the fault at F1, 89–96
  solution assumptions in, 66–67
  steps for, 65
Transient stability problem in simple electrical network:
  electric power transmitted, 53–54
  network reduction, 50–52
  numerical solver, 58–64
  power transmitted before, during and after fault conditions, 55–56
  stability problem, 49–50
  swing equation, 56–58
Transmission:
  lossless short line, 113f
  parallel lines, 40, 49, 50f
  three-phase balanced circuit for, 1
TRVs. *See* Transient recovery voltage capacitors
Two-layer earth model:
  ground grid buried in bottom layer, 206–207, 206f
  ground grid buried in top layer of, 203–205, 204f
  ground mat design principles, 159
Two-phase to ground short circuit failure, 1

U

Utility contribution to short circuit MVA, 11

V

Ventricular fibrillation of heart, 132, 134
V_{mesh}. *See* Mesh voltage

Index

Voltage. *See also* High-voltage AC capacitors
  bus, before completion of first iteration, 81–82
  capacitor, 103–125
  E_{step}, 134, 135f, 138, 148
  E_{touch}, 128, 135, 136f, 138, 148
  generated, 20
  generator-infinity bus network system, 39–40
  generator terminal, 22
  let-go, 136f
  LL, 5, 72
  LN, 5, 71, 113, 116, 124f
  V_{mesh}, 138, 148, 149, 151f, 197–203, 203f
  oscillations during first four seconds, 123–125
  output of synchronous generators, 17
  sinusoidal instantaneous midpoint, 115–119
  V_{step}, 138, 148
  swing bus magnitude, 81, 83
  V_{touch}, 138, 148
  V_{trans}, 138, 149

Voltage (*Cont.*):
  TRVs, 104
  winding application of, 13f
  V_{step}. *See* Step voltage
  V_{touch}. *See* Touch voltage
  V_{trans}. *See* Transferred voltage

X

X-directed segment of point current source, 164–165, 198–201
  self-resistance, 174
  two parallel, 168–169

Y

Y-directed segment of point current source, 164, 166, 199–202
  parallel, 169–170
  self-resistance of, 174

Z

Z-directed segment of point current source, 165, 167
  parallel, 170
  self-resistance of, 175